# GROW YOUR OWN PLANTS

*From Seeds, Cuttings, Division, Layering, and Grafting*

# GROW YOUR OWN PLANTS

*From Seeds, Cuttings, Division, Layering, and Grafting*

*by* JACK KRAMER

*Drawings by Michael Valdez*

CHARLES SCRIBNER'S SONS · *New York*

1 3 5 7 9 11 13 15 17 19 MD/C 20 18 16 14 12 10 8 6 4 2
1 3 5 7 9 11 13 15 17 19 MD/P 20 18 16 14 12 10 8 6 4 2

Printed in the United States of America
Library of Congress Catalog Card Number 72-7939
SBN 684-13233-8 (cloth)
SBN 684-13234-6 (paper)

BOOKS BY JACK KRAMER

*Hanging Gardens*
*Water Gardening*
*Miniature Plants Indoors and Out*
*Garden Planning for the Small Property*
*Gardening with Stone and Sand*
*Natural Dyes: Plants and Processes*
*The Natural Way to Pest-free Gardening*
*Ferns and Palms for Indoor Decoration*
*Your City Garden*
*Grow Your Own Plants*

# Contents

# *Introduction: A World of Plants*

If you have ever said you cannot grow plants you have a surprise waiting for you when you grow your own from seeds, or from cuttings, by division, air layering, and grafting. You need no elaborate green thumb to bring plants into being. Seeds are like magic and provide a wonder of nature as you watch your plants grow to perfection. Taking cuttings from plants is a simple procedure and so is dividing plants to make new ones. Air layering and grafting of plants are other ways to increase your garden stock.

Growing your own plants is not instant gardening; it takes time and patience, and planning, but the rewards are vast—a head start on spring, a selection of plants not otherwise available, and best of all, generally sturdier plants than you can buy. And too, let us not forget the economy of growing your own; you do save money.

In addition to all this, there is something more. To watch the mystery of life as it develops is fascinating, even to the non-gardener. The miracle of growth is not only pleasing to the eye but to the soul as well. This is a rewarding hobby that even children can practice with success. Nature is full of magic, and plant propagation is only one part of her bag of tricks.

In this book, we hope to tell you simply and clearly, with as few technical terms as possible, how this magic works. We start at the beginning with seeds for they are the essence of life. We show and tell you the various steps of many kinds of plant propagation and how you can grow almost any kind of plant—tree, shrub, perennial, annual, vegetable, herb, and bulb—from start to finish.

*Jack Kramer*

# 1. The Miracle of Plant Growth

The word propagation sounds technical, and some of the processes of propagation—air layering, grafting, division, and seed sowing—may seem complex and mysterious. But propagation is simply a way of increasing plants. In many cases propagation is so simple that even a child can do it; what complicates this simple procedure is the many *methods*. Each method (sowing seed, layering, grafting, division, and cuttings) is explained in the following chapters, and each has a definite use.

Propagation is basically from seed (sexual) and vegetative parts of a plant (asexual); following are some elementary remarks on these two types of reproduction.

## Sexual Propagation

Flowers are lovely and a joy to look at, but their ultimate purpose is seed production, to perpetuate the species. Propagation by seed, the most common means, is also the readiest and least expensive. This sexual method produces seedlings that will vary somewhat in their characteristics since seeds are not exact reproductions of their parents. But these seedlings may adapt better to survival.

Seeds are actually fertilized embryos that when mature include rudimentary plants, protected seed coats, and internal nutrients to supply growth. A seed is made up of an embryo, seed coat, and sometimes endosperm, and is produced by a flower. Two organs in flowers produce a seed: the stamen, which has pollen grains that eventually form the male cells, and the pistil or female organ, which is generally in the center of the flower.

# METHODS OF PROPAGATING PLANTS

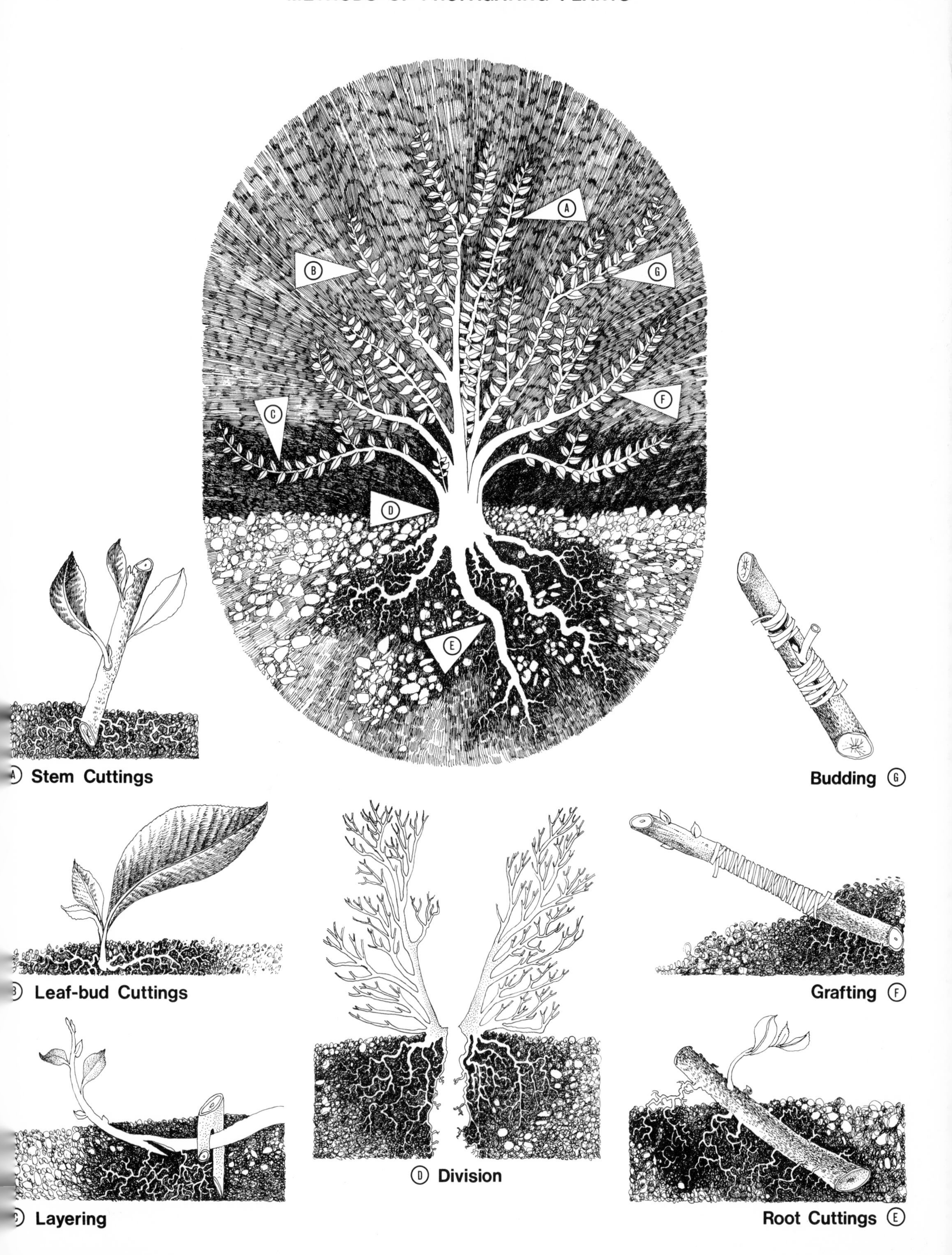

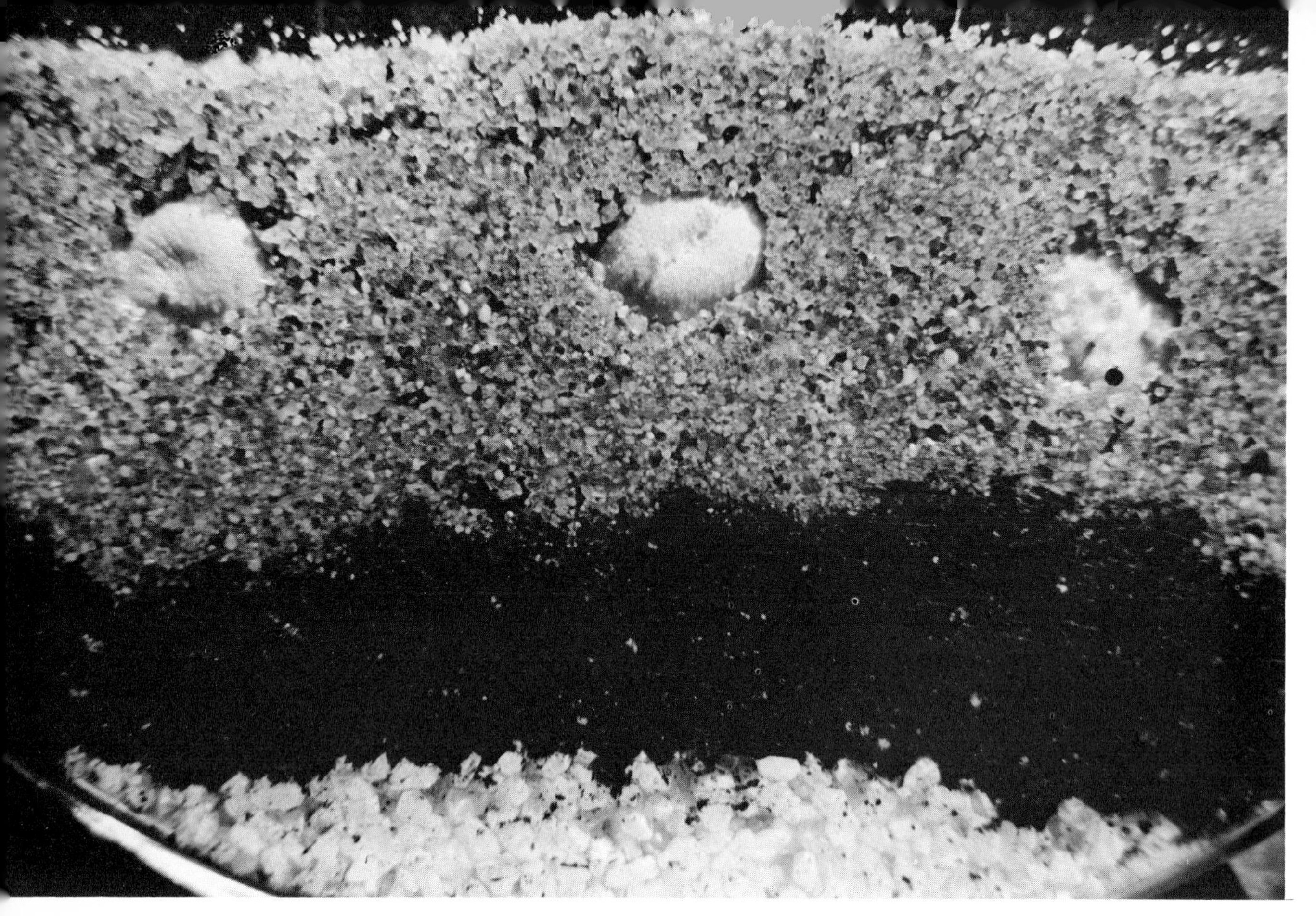

*The miracle of plant growth is well shown in these photos. Above, lima bean seed has been sown in growing medium. Below, roots and leaves are evident and a new plant is born.* (Photos by J. Barnich)

The steps leading to formation of seeds are:

*1. Formation of stamen and pistil*

*2. Opening of flower*

*3. Transfer of pollen from stamen to pistil*

*4. Germination of pollen*

*5. Fertilization*

*6. Growth of the fertilized egg into an embryo with a surrounding case, the seed*

The word miracle aptly describes a seed. A seed is a veritable powerhouse of stored food that can, if necessary, survive freezing and drought. When we garden with seeds we are imitating nature and thus have her same problems: providing the right conditions—temperature and moisture—so seeds will germinate and grow.

Seeds come in many sizes, from the fine dust of some begonias to the pea-sized seeds of morning glories. Some seeds can catch the air and ride with the wind or, as is the case with spruces, develop a sort of catapult device that snaps them into the air a hundred feet or more. Some seeds like dandelion and milkweed develop a flimsy parachute of silky hairs; you have no doubt seen these tufts of hair floating on the wind for thousands of feet. And some seeds, e.g. lotus, have a watertight protective coating that enables them to float for miles until they find a suitable place to germinate. (Birds and animals also transport various seeds.)

What makes seeds start to grow? Different degrees or quantities of viability, water, air, temperature, and age and stage of maturity produce the proper life-starting combination. Heat may cook seeds, and cold, along with frost, may injure some seeds. Thus, special conditions are necessary for germination: proper humidity, ventilation, light, and moisture. The stages of germination involve absorption of moisture by the seed, favorable temperatures to transform stored food into sugars by enzymes or natural ferments, and the bursting of the seed coat.

A seed needs water because the plant foods it holds must be in solution if they are to be of any use to the embryo. Too little moisture or not enough will fail to do the trick, but an evenly moist soil that provides moderate moisture starts the germination process. Light also affects germination; some seeds are indifferent to light exposure, but others need light to start growth.

*Seed pods come in different sizes and shapes; here we show (left to right) Lupine, Oriental poppy, and Allium.* (Photo by author)

*Seeds, too, come in many sizes. On the left, the small black seed of Allium; in the center, the microscopic seed of the poppy; on the right, the relatively large seed of Lupine.* (Photo by author)

*Commercial seed is available in packages ready for planting at nurseries and plant centers.* (Photo by author)

*Seed shapes and sizes vary from minute to large.* (Photo by author)

*In the flower nursery of Burpee Seed Co. we see plants in full bloom ready for pollination.* (Photo courtesy Burpee Seed Co.)

# SEEDLINGS

AGERATUM
A. houstonianum

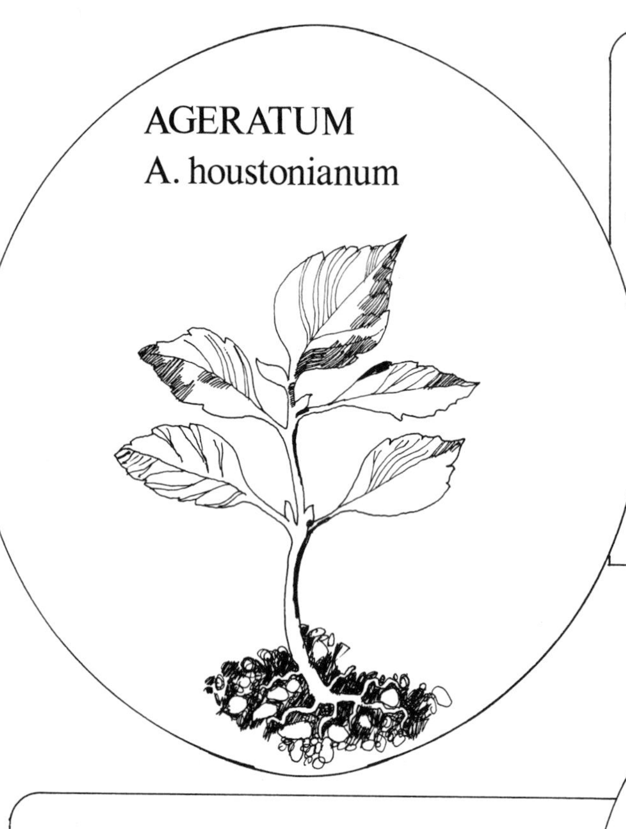

WAX BEGONIA B. semperflorens

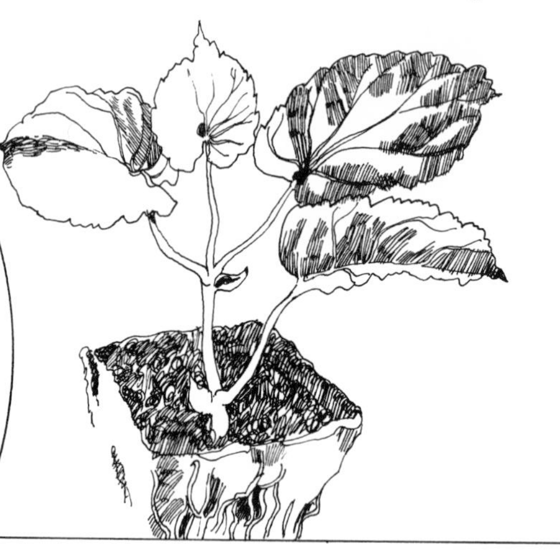

BALSAM
Impatiens
balsamina

CANTERBURY BELLS
Campanula medium

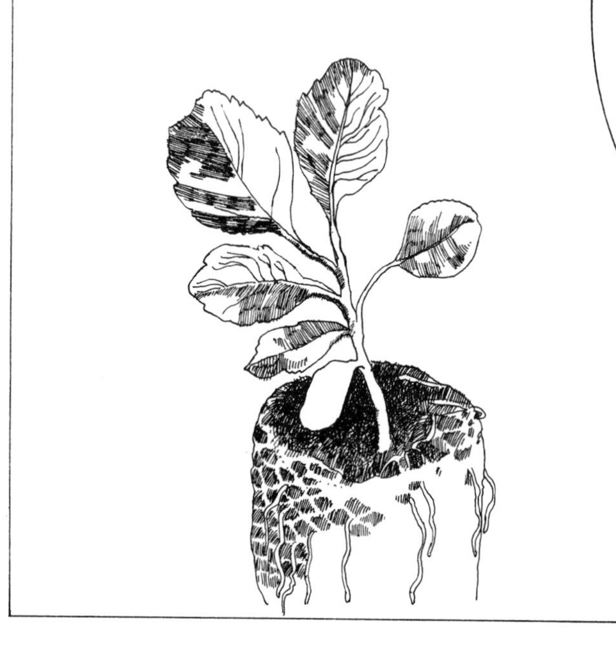

PAINTED DAISY
Chrysanthemum coccineum

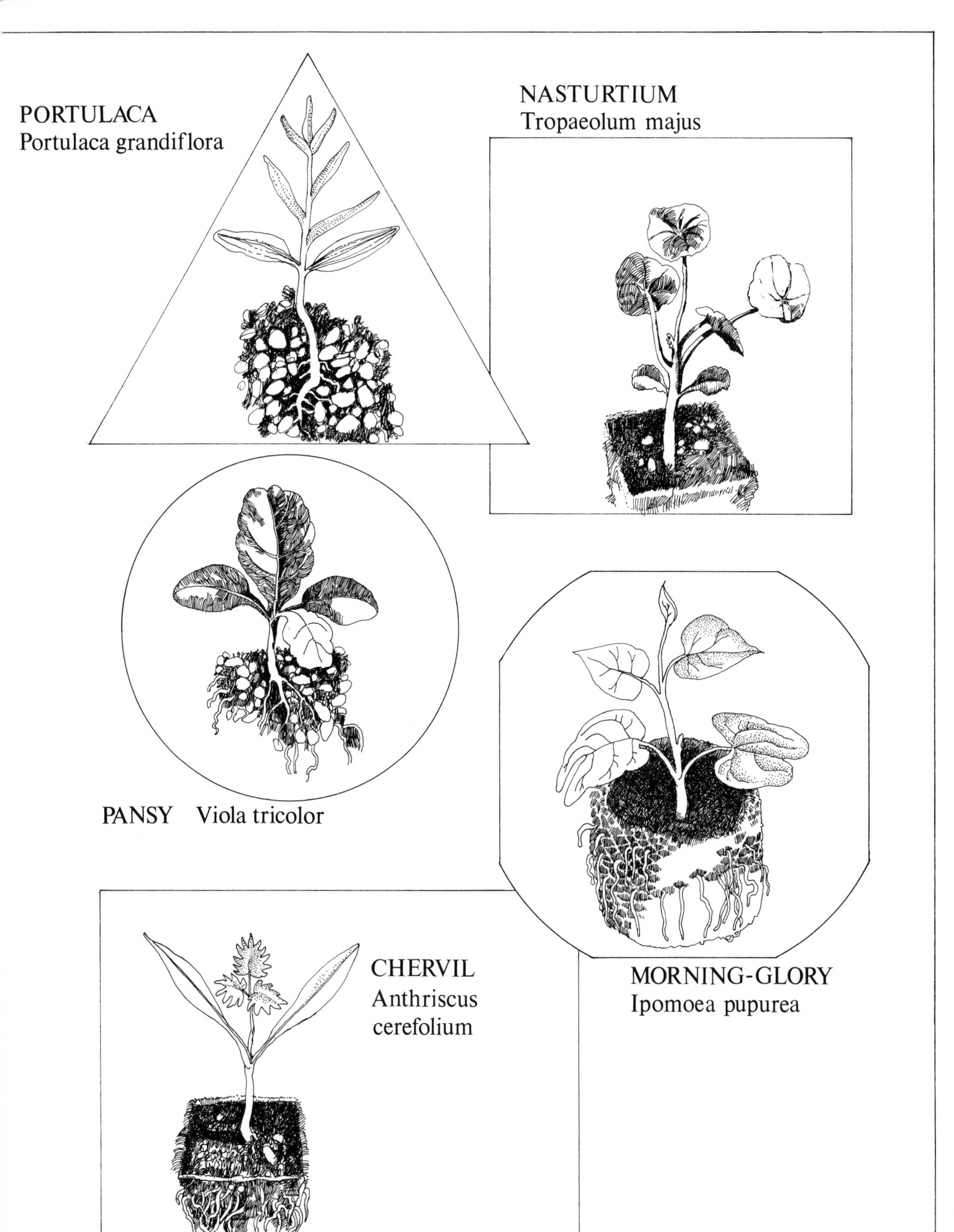
PORTULACA
Portulaca grandiflora
NASTURTIUM
Tropaeolum majus
PANSY Viola tricolor
MORNING-GLORY
Ipomoea pupurea
CHERVIL
Anthriscus
cerefolium

## SEEDLINGS

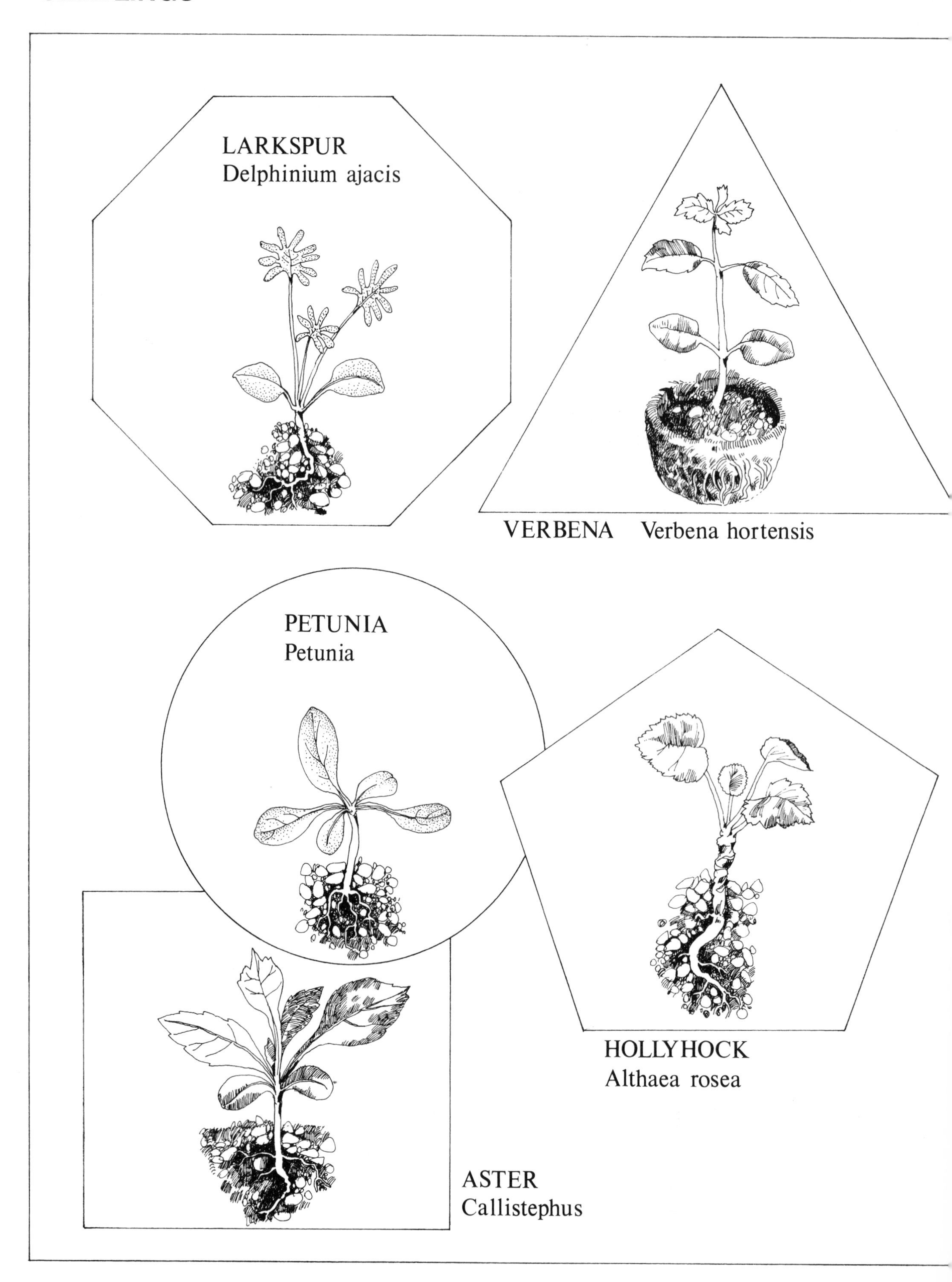

LARKSPUR
Delphinium ajacis
VERBENA Verbena hortensis
PETUNIA
Petunia
HOLLYHOCK
Althaea rosea
ASTER
Callistephus

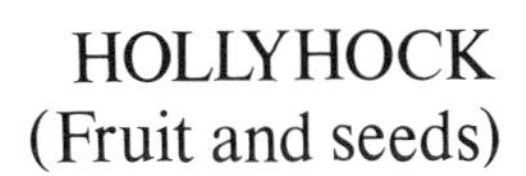

HOLLYHOCK
(Fruit and seeds)

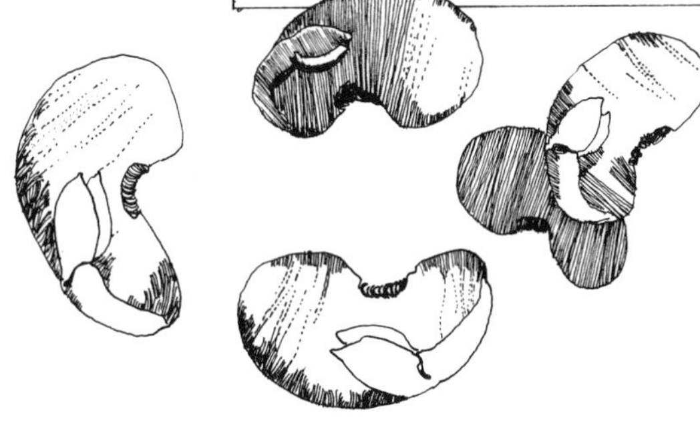

HOP HORNBEAM

HAWTHORN

ASH SEEDS

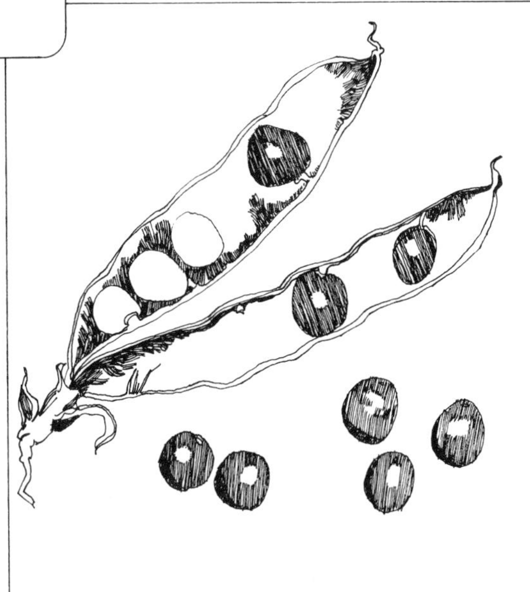

BEAN POD

# SEEDS

WALNUT

SUGAR MAPLE

PECAN

BURR and RED OAK

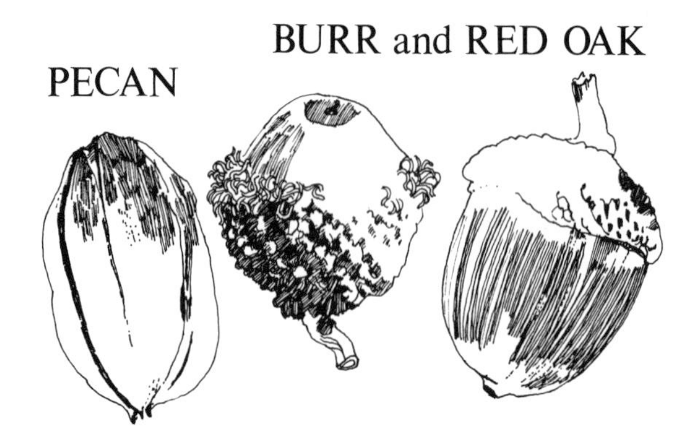

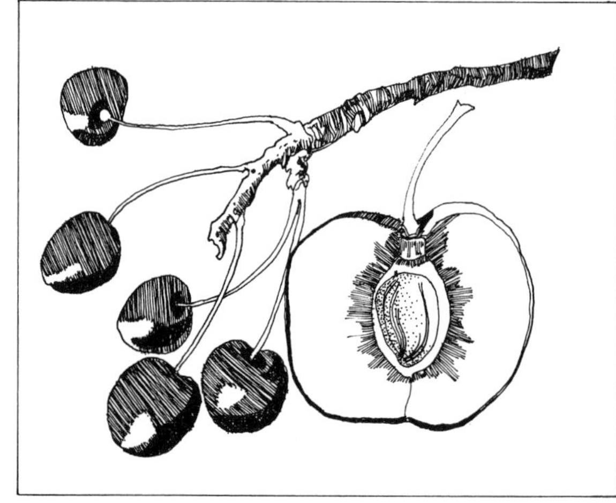

### Commercial Seeds—Your Own Seeds

Most gardeners buy seeds already processed and packaged from suppliers. But of course you can take seeds from your own plants. However, these seeds must be taken from true species; seeds from hybrids (best crossed with best), which are part of most gardens, will not produce plants exactly like their parent. They may revert to an unsatisfactory characteristic of one of the parents used in developing the hybrid, such as smaller flowers or muted colors. Still, collecting seeds offers adventure—the lure of the unknown. Generally, when you collect your own seeds, do it when they are ripe, and then clean and store them in a cool place until you sow them. Some types, however, have to be gathered before they are fully ripe to prevent dispersal. With pines, for example, fully matured seed cones can open rapidly in a warm spring day and then seed is lost. When pollinating flowers be sure that cross pollination doesn't occur (see Chapter 2).

No further processing or care is needed with commercial seeds. They are ready for use when you get them. Commercial seed growers constantly seek new ways to bring the very best plant strains to the public. The commercial plant breeder knows exactly the proper stage of maturity for harvesting, the proper seed-collecting methods, the proper cleaning and processing means, and most importantly, the proper way of storing seeds so they remain viable until they are used. Within each group of plants—perennials, woody plants, annuals—there are special considerations; we shall discuss them in other chapters.

### Asexual Propagation

You have probably unknowingly practiced this method of increasing your plants. Have you ever taken a runner from a house plant like chlorophytum or *Saxifraga sarmentosa,* cut it from the parent plant, and potted it in soil? Have you ever, when repotting a plant too large for its container, taken a chunk and started it growing separately? These are all asexual propagation methods, that is, using such vegetative parts of the plant as roots, stems, or leaves to get new plants. With this type of propagation all the characteristics of the parent are passed on to the new plant because an exact duplication of the chromosome system occurs during cell division.

*Taking cuttings of plants is one way of asexual propagation.* (USDA photo)

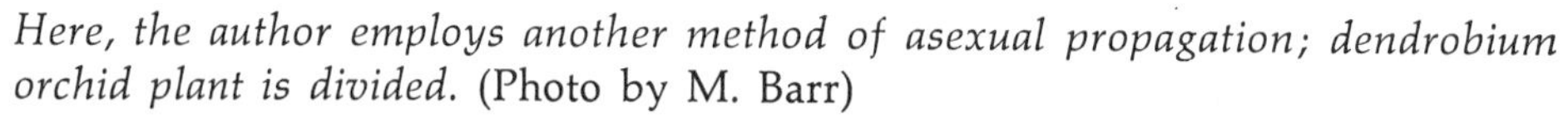

*Here, the author employs another method of asexual propagation; dendrobium orchid plant is divided.* (Photo by M. Barr)

Propagating cuttings is perhaps the most convenient way of getting new plants for your garden (a cutting generally refers to the stem but may also be a leaf or root cutting). Starting a cutting can be done in a few minutes, does not harm the parent plant, and is ridiculously simple. But not all plants can be satisfactorily increased by cuttings; most stone fruits must be grafted or budded because they do not strike root from cuttings. The chief advantages of cuttings instead of seeds is that often you get a stronger plant in less time and, as mentioned, the new plant is a duplicate of the parent plant.

For the indoor gardener who wants more plants, leaf propagation, i.e. taking a piece of leaf from the plant and starting it in a propagating mix, is a good method. African violets and begonias can be started in this manner.

After cuttings, runners, leaf cuttings, or plant division, proper care is necessary to nurture the new plants. Like seeds, they require a growing medium, good humidity, adequate ventilation, and even moisture. When roots form and plantlets start growing, transplant them in soil in individual containers.

SUMMARY OF THE METHODS OF PLANT PROPAGATION

1. *Sexual*
   - Seed
     - Sowing
     - Germination
     - Thinning
     - Transplanting
2. *Asexual division*
   - Offsets
   - Crowns
   - Tubers
   - Rhizomes
   - Runners
3. *Cuttings*
   - Stem
     - Hardwood
     - Semihardwood
     - Softwood
     - Herbaceous
   - Root
   - Leaf
4. *Grafting*
   - Root
   - Crown
   - Top
5. *Layering*
   - Simple
   - Tip
   - Mound
   - Trench
   - Air

# 2. Seeds Indoors

Growing your own plants from seeds may seem anachronistic in these days of speed and convenience, but plants you raise from seeds bring you immense satisfaction, and if you are gardening even on a moderate scale they save you money.

No matter which seed-starting media you use—and there are many—it must be porous so air and water can reach the seed.

At first, sprouting seeds need no food. Photosynthesis does not start until there are true leaves, so do not use fertilizer until it is actually needed. But when the first true leaves (cotyledons) appear, food is necessary. The main thing is to furnish a constant supply of moisture and proper temperatures so seeds can sprout. In other words, be careful and precise about getting seeds started; give them your attention and care.

As mentioned, you can obtain seeds from your own plants and trees or obtain them from supply houses. Once gathered, seeds from your own plants should be removed from cones or pods and prepared either for storing or sowing. To store seeds be sure they are dry and clean and remove any pulp; how you store seeds makes a difference in their viability. (The viability of the seed depends on the variety.) When properly stored, many seeds remain good for a long time. Put seeds in an airtight, covered jar or container and place them in a cool, dark place. Keep seeds dry, and see that they are not subject to temperature fluctuation.

### Germination

Some seeds are difficult to germinate because the coatings are so watertight—nature has done her job well—that germination is

*Peat pots being filled with soil for seed sowing to be placed in cold frame. When transplanted, pot and plant go into the ground.* (Photo by author)

delayed dangerously. Such seeds can be soaked overnight in water or, in extreme cases, in dilute sulfuric acid (handle with care) before planting. Other seeds will need stratification or scarification.

*Stratification*

There are variations in the germination of many seeds. The process may be delayed if a period of rest, which is necessary in the normal germination sequence of some seeds (especially woody tree and shrub seeds), is not provided. For these seeds, a process known as stratification is sometimes used. The seeds require moisture, oxygen, and a relatively cool temperature similar to what they get naturally in winter months in the forest. To simulate these conditions, provide temperatures of 32° to 50° F. for 1 to 4 months depending on the type of seed being sown.

Prepare seeds for this process by soaking them in water for 12 to 24 hours; then put them into a moisture-retentive medium. Suitable materials include sand, peat moss, chopped sphagnum, and vermiculite. Mix seeds with the medium or place them in layers 1 to 3 inches thick, alternating with equal layers of medium in containers outdoors (boxes, cans, or glass jars can be used).

When the afterripening period (1 to 4 months) is over, seeds may show signs of germination. Remove them then from the medium with a screen mesh and immediately plant them.

*Seed being sown in peat pots in cold frame.* (Photo by author)

*Peat pellets are also used for seed sowing; various stages of development to "pot" with petunia growing in it is shown.* (Photo courtesy George W. Park Seed Co.)

It is senseless for the home gardener to make special boxes, with layers of peat, sand, or peat-soil mixtures, for outdoor stratification. Simply mix some shredded sphagnum moss with pre-soaked (12 hours) seeds, place them in a plastic Baggie, and store the mixture in the refrigerator. When seeds start to germinate in the stratification medium, separate them from the medium and plant.

### *Scarification*

For tough-coated seeds use the process of scarification, which is nothing more than nicking the seed with a knife or other sharp instrument. Once the coating is permeated, water can enter the seed, and if all other conditions are good, germination is more likely to occur. *Cobea scandens* is a typical example of a seed that needs scarification.

### *Soaking Seeds*

Soaking seeds in hot water softens seed coats and reduces the germination period. Put the seeds into hot water (180° to 210° F.), and allow them to soak for 24 hours as the water cools. Do not boil the seeds because this is injurious to most seeds. Sow seeds immediately afterward; if they dry, the treatment is ineffective (but overly wet seeds are no good for sowing either). The soaking process is simple, inexpensive, needs no special equipment, and is very effective for a number of different kinds of seeds. Soaking seeds in sulfuric acid is another way to hasten germination, but I do not suggest it for the home gardener. Sulfuric acid is strongly corrosive, and dangerous to use for the novice.

## Advantages of Seed Growing

The gardener who grows his plants from seeds garners more than just economy. He can have the latest introductions of plants, sometimes in the first year. New introductions of both annuals and perennials take years to get from originator to public testing to general nursery propagation. I still remember the joy of having the first dwarf red impatiens in full bloom in planter boxes in my garden long before they were available to the public.

Because of the increasing production of new and lovely hybrids, some plants are lost in the shuffle; it is these forgotten favorites we

*Seed sowing helpers include pellets, peat pots, cubes . . . all available from suppliers* (Photo courtesy George W. Park Seed Co.)

occasionally find in friends' gardens. A friend gave me some seeds of a forgotten species of iris that had been growing in her garden for fifty years but was unavailable commercially. If you can get such seeds you can preserve beautiful but "rare" species.

Many times the color of flower or variety of vegetable that you want for your garden is not available at your local garden center. Growing plants from seeds makes it possible for you to have what you really want—and not what you must buy.

Getting a head start on spring is another bonus for the seed grower. He can start many of his plants indoors and have them in better health at planting time than store-bought seedlings. Roots will be robust, and plants will be really ready to grow, once set in the garden. Furthermore, even though many fine plants are sold at suppliers, the gardener who grows his own seeds always gets more for his money and a sense of satisfaction. There is immense reward in nurturing a tiny seed into a perfect plant.

Expensive or elaborate equipment is not necessary for growing seeds, nor is a knowledge of botany; all that is required is a love of nature and a certain curiosity that is part of everyone's makeup.

### Seed Containers

There are dozens of seed containers, some professionally made and others homemade. Suppliers carry starting kits, jiffy pots, seed cubes, plastic-bubble seed containers, and so forth. Just what you use depends on your own personal taste. For years, ever since I worked for a glass factory, and before commercial seed kits were available, I used the boxes from the factory as containers. Indeed, commercial nurseries still use these. These containers range from 12 x 18 inches to 24 x 30 inches in size. They are 3 to 4 inches deep and have spaces between the bottom boards for drainage. These boxes, or "flats," as they are sometimes called, are still available free from glass dealers in some cities.

Household items also make good seed containers—for example, coffee cans, aluminum and glass baking dishes, plastic cheese containers, and aluminum pans that frozen rolls come in. Be sure that any household item you use as a container is at least 3 inches deep and has some drainage facilities. Punch tiny holes in the bottoms of the containers. Also remember that the planting mix in aluminum and plastic housings will dry out quickly, so more frequent watering and attention is necessary. Generally, a plastic covering is used to assure good humidity.

I suggest that the beginner use the standard clay pots sold at nurseries. They are inexpensive, always look neat, and hold moisture, so frequent watering is not necessary. Ask for azalea pans—squatty pots—which are now available in several sizes.

Other suitable containers include galvanized metal pans (sometimes found in used-furniture shops) or custom-made ones available at metal shops.

*Here we see seed cubes, yet another way to start seeds.* (Photo courtesy George W. Park Seed Co.)

# SOWING SEED IN FLATS

① Fill with sterile growing medium and moisten.

② Sow seeds according to direction on packet. Cover seeds lightly with s

③ Cut dowels to lengths of one-half inch from top of flat; insert dowels in corners on one side of flat. Top with glass panel or plastic covering to insure humidity.

④ Remove seedlings for transplanting.

Seed-growing Mediums

Whether you put seeds in flats or pots, a suitable growing medium is essential for good germination. There are many kinds of mixes for growing seeds: peat, perlite, sand, and vermiculite. But no matter which one you use it should be firm and dense, sterile, able to retain moisture (but not too much), and, most importantly, sufficiently porous so it allows good water drainage. The ultimate choice of the mix depends on the plants you grow. We find that a peat-like mix is excellent for cacti and succulents, but we use standard vermiculite and milled sphagnum for annuals and perennials. For trees and shrubs we use equal parts of sphagnum, perlite, peat, and vermiculite.

You can prepare your own mix from the following materials or buy commercially packaged kinds like Jiffy-Mix, Pro-Mix, etc. In any case, use a *sterile* medium or the soil-borne fungus known as damping off can occur. If it does, you will have weeds, nematodes, and various disease organisms.

*Seeds being sown in wooden flats; don't forget to label flats so you will know what you are growing.* (Photo by author)

*You can also start seed in this setup . . . a pot within a pot. Vermiculite is the growing medium.* (Photo by author)

You can use regular soil, but most commercial growers start seeds in various mixtures of peat, vermiculite, soil, perlite, and sand because they are uniform, sterile, and hold moisture well. And it is easy to remove seeds at transplanting times without injuring roots.

The following growing mediums can be used for tray- or pan-seed sowing under lights, on windowsills, or in basements and similar indoor areas. For open-garden or cold-frame sowing see Chapter 3.

*Perlite.* This is a volcanic ash that does not absorb moisture readily but holds it within itself, providing a moist growing medium. Mixed with a little sterilized soil it provides a good medium; by itself it has a tendency to float and disturb the seed bed.

*Milled sphagnum.* This is an old-timer and a good starting medium that generally gives good results. However, it has to be carefully watered to keep an evenly moist bed.

*Vermiculite.* This is expanded mica that holds moisture a long time; we use it with excellent results. It is sold under various trade names and is packaged with added ingredients.

*Packaged soil mixes.* Steer clear of these if at all possible because they are generally too heavy for successful seed sowing. But if nothing else is at hand, use in combination with some finely porous sterile soil.

*Sand.* Use a fine quartz sand with a silica content. Be sure that the medium is always slightly moist but never soggy. Sand has a tendency to become too dense with excessive moisture and retards germination. It is best to use sand in combination with soil and vermiculite.

*Peat.* Peat is made of the remains of marsh, bog, or swamp vegetation in a partially decomposed state. Peat varies widely in quality and kind. The light brown or yellow-brown types are generally quite acid in reaction; the brown or black types are better choices. Peat holds water well, and has some nitrogen but is low in phosphorus and potassium. Before adding to mix, break peat apart and moisten it.

*Soils.* An ideal soil mixture should have porosity so air and water can enter and reach seeds. It should be free from harmful organisms (be sterilized) and crumbly in texture.

### Sowing Seeds Indoors

To prepare small containers such as azalea pots, coffee cans, and aluminum trays, fill them to within a half inch from the top with a sterile, moist mix. Press the material in place to eliminate air spaces. Sow seeds about half an inch apart; cover large and medium seeds with a layer of dry mix (twice the thickness of the seed is the general procedure, for some seeds like verbena, pansy, and larkspur need a

3- to 5-day period of darkness for germination). Merely scatter other seeds, such as impatiens, and begonia, on the top of the growing medium. Sprinkle the surface with water. Make a tent of plastic bags on sticks with a few air holes punched in the plastic and cover the container.

The procedure is the same with large containers such as flats and trays, except that more seeds are used to fill the containers.

Place planted containers in a bright (but no sun) area with temperatures of 65° to 75° F. by day. At night a drop of 10° is fine. Keep the seed bed moist but never wet.

As mentioned, water, temperature, and air are necessary for germination. The amount of moisture affects both the percentage of germination and the germination rate. Once sprouted, seedlings will die if they become dry, so uniform moisture is of utmost importance. Not enough water will cause plants to die, but too much water reduces aeration in the medium and leaves plants prone to damping off.

Some growers recommend pre-soaking seeds to shorten the time for seedlings to sprout. However, except for seeds with excessively hard shells, we have found little difference in germination time between pre-soaked and nonsoaked seed.

*Household containers can also be used for seed sowing; here a refrigerator dish and a plastic box are the containers. Lids are used to provide humidity for the seeds.* (Photo by author)

Seeds require a favorable temperature to germinate, and although there is a wide range of temperatures, most seeds do need heat, about 72° F. Some germinate only within a certain narrow range. Invariably, warmer temperatures produce more plants. Germination is often better if seeds have alternating rather than constant temperatures; normally, nighttime temperatures are lower than daytime (and this should be encouraged).

### Care of Seedlings

Check the growing medium daily to be sure it is uniformly moist on top as well as on the bottom. When the second set of leaves emerges, move containers to bright light and remove all plastic or glass coverings. Keep the medium uniformly moist; weak fertilizer such as Hyponex or Rapidgrow can be used once every two weeks now. When seedlings are about one inch high and have separate leaves, transplant them to pots with regular soil.

To remove the seedlings, lift them out carefully with a blunt-nosed stick or your finger, capturing as much of the rootball as possible. Place seedlings in small pots, or put them in flats and water well with a fine-mist spray. Give them full light now. Gradually acclimate seedlings to outdoor conditions before planting them outdoors. (This process is known as hardening off.) Put seedlings in a protected place—near a house wall, or under shrubs—where they will be out of wind and very bright sun. Gradually expose seedlings to outdoor conditions. Be sure the soil does not dry out, and if frost threatens, return the plants to indoor locations.

When plants are properly hardened (dark green leaves) and all danger of frost is past, put plants in permanent places in the garden. Once again, remove them with as much of the rootball intact as possible, and after transplanting be sure to water them thoroughly. Continue light feedings as plants mature.

### Propagation Helpers

Even if you are only taking cuttings from house plants, you should be aware of rooting hormones, since they are great helpers. These synthetic plant hormones, available under the trade names Rootone, Hormodin, etc., promote the rooting of stem cuttings. The hormones

*An excellent propagation helper is root hormone powder; cuttings are dipped into powder to stimulate root growth.* (USDA photo)

induce or stimulate the growth of roots on cuttings. (Various chemicals such as indolebutyric acid are also used, mainly by commercial growers, for treating cuttings.)

It is a simple matter to apply the powdered root-inducing hormones to plants. Dip the base of the cutting about a half to one inch in water and then into the powdered hormones. Many cuttings are succulent and watery at ends and will not have to be moistened first. Do not douse the end of the cutting with too much powder; just be sure it is fully coated. Then follow the usual planting steps.

It is not mandatory to use Rootone or other commercial products on many plants, but it has been found that the majority of plants respond by producing more roots and in less time. Do not, however, expect root-inducing hormones to be miracle workers; cuttings will still need the proper care to prosper.

## VARIOUS METHODS FOR PROPAGATING CUTTINGS

Use glass jar for single cutting.

Make plastic tent for few cuttings.

Use aquarium or plastic
frigerator dish for multiple cuttings.

Polyethylene plastics do more than store foods. They help you in many ways, and perhaps the most important way is with plants. We started using plastics some years ago when we wanted to provide additional humidity for plants. We simply covered the plant with the plastic, and propagation was greatly enhanced.

This flexible plastic—available as Baggies and so forth—retains moisture while admitting light. Thus it is a very usable material for making small greenhouses. Place wire or wood sticks in a flat and cover with plastic. The plastic, as mentioned, helps to provide adequate moisture. You can also drape plastic over a clay pot of seed or simply wrap cuttings in it. Place some moist sphagnum moss in the bag of plastic, and, after dipping the cutting in Rootone, place it on the moss, with the plastic tied at each end. When roots appear the cutting can be removed from its protective housing and planted. Be sure that you don't get too much humidity when using polyethylene plastic. If the plastic becomes too moist (and you will see this), remove it for a few hours a day or poke small holes in it.

# 3. Seeds Outdoors

As explained in the previous chapter, seeds can be started indoors. However, if many plants are needed, a cold frame or hotbed outdoors is an essential piece of equipment. These seldom-seen units are rarely publicized because they are not glamorous or attractive. Yet they are invaluable to the gardener who wants to grow his own plants from seeds and cuttings. The structures can be tucked into the smallest part of the garden; an area 5 x 10 feet will accommodate the standard 3 x 6-foot cold frame.

Just what are cold frames and hotbeds? Basically they are miniature greenhouses: four wooden sides with a glass or plastic covering. Cold frames use the heat of the sun; the hotbed (similar to the cold frame) has provisions for artificial heat and can be used all year.

### Cold Frames

The addition of a cold frame is a real convenience to even the beginning gardener. It is a helping hand, a place to store bulbs and keep plants for hardening off, and, of course, a place to grow under good conditions a variety of plants from seeds or cuttings. It provides a place to stratify seeds, too.

The cold frame can be of any size, but it is generally a wood box or frame that has sloping sides with a top of glass or hotbed sash. (Regular window sash or plastic are satisfactory too.) Because hotbed sash is available in 3 x 6-foot sizes, most cold frames are built to this dimension, or multiples.

The cold frame can be built from 2-inch redwood with 4 x 4-inch corner posts. The structure is generally bottomless, the frames sunk 2 to 3 inches in soil. The sash may rest on the top of the boxes or be hinged at one end for convenience (you should be able to lift the sash at one end to provide ventilation). We use small blocks of wood to prop our panels, which are of rigid plastic. We find it easier to lift at will and have discovered that the translucent quality of the plastic allows enough, but not too much, light for plants. Thus we don't have to worry about shading the plants.

Locate the cold frame so its length is on a north-south line and in full sun; be sure you have space around it so you can lift the tops when necessary to tend the seedlings. I made the mistake of setting my cold frames too close to each other and it was difficult to take care of the plants.

Once the frame is in place, fill it with 6 to 10 inches of porous soil. Rake and tamp it down and then moisten the soil with a light spray. It is then ready for seed sowing. You can also set flats of seed in the cold frame or simply put in peat pots as we do without an additional soil bed. Once seeds are in place, never let the soil bed (or peat pots) dry out. Even moisture is essential, using a fine spray of water; so is ventilation and shading. And don't forget to identify plants with markers—plastic or wood. Or simply tack the seed pack on the cold frame wall.

*These cold frames are simple to make; redwood boards were used.* (Photo by author)

*Peat pots with seed in cold frame.* (Photo by author)

*The author's cold frames with seedlings. Note the translucent plastic on left used to protect seedlings from intense sun.* (Photo by author)

*Water seedlings with light mist; a strong stream of water will dislodge seeds.* (Photo by author)

*This homemade cold frame uses discarded window sash for a top.* (Photo by R. Tracy)

When seeds are first sown, keep the top closed to ensure good humidity. As germination takes place, partially raise the sash to provide more ventilation and drier conditions so damping off does not occur. In sunny weather ventilation is of prime importance for heat can build up quickly in the cold frame and desiccate young plants. Shading too, will, in most cases, be necessary. You can use wood laths on a frame or cheese cloth stretched over a frame, paint the glass with a shading paint or use translucent plastic as I do to protect the seedlings from intense sun.

Commercial cold frames are now available (although expensive) from garden suppliers. These are generally metal units and can certainly be used if you do not want to build your own cold frame.

### Hotbed Construction

As mentioned, hotbeds are constructed in the same manner as cold frames, but they need a source of artificial heat. Electric soil-heating cables, sold at suppliers, are satisfactory and in most cases (unless you are handy) should be installed by professional electricians. Be sure a thermostat is included so exact temperatures can be maintained. Place soil over the heating cables or, if you merely want to set flats of seeds on the cables, install a 1-inch layer of sand first.

The more complicated method of heating the hotbed (often suggested in gardening books) involves electric light bulbs in sockets that are placed at varying distances. Don't bother with this construction. It is a mess to put together and to keep operating properly. Soil cables are better by far.

*Ventilation within a cold frame is of vital importance so damping-off of seeds does not occur. A simple wooden block can be used to prop open sash.* (Photo by R. Tracy)

### Sowing Seeds Outdoors

If you don't want to tend a cold frame or bother with indoor-container seed sowing, start annual, perennial, and vegetable seeds directly by planting in the open garden. However, this type of sowing is not as dependable as the cold frame method. Outdoor sowing depends upon proper timing—too early or too late and the results are disastrous. If weather is still cool, sow only those varieties that are noted for being cool plants (among the vegetables this includes peas, carrots, beets, and lettuce). On the other hand, some plants, including many of the best perennials, must have warmth to germinate properly.

Soil is of utmost importance for sowing seeds directly in the garden. It must be good soil, full of nutrients, so plants can grow. It is true that a plant like nasturtium will germinate in practically any soil, but most plants need a rich friable soil. Thus you will have to till or fork the soil over first; do this when soil is dry. Wet soil will become a sticky lumpy mess, difficult to handle.

Here in California we dig to a depth of 18 inches as we work and turn the soil. A garden tiller can, I suppose, be used, but we generally rely on a spade. Once the soil is turned and crumbly, we add compost and leaf mold. This will condition the soil and help bacterial growth, which produces decomposition, so plant nutrients are available to the plants. Mix the compost and loam thoroughly (how much of each you add depends on the plot of ground you are cultivating). By the way, all ingredients mentioned are now available in tidy packages. We use equal parts of compost and leaf mold with existing soil. At this point it is wise to test your soil, to see if there are any deficiencies that can be corrected by supplemental feeding.

Soil is tested according to a pH scale. On the pH scale 7.0 is neutral; the soil is neither acid nor alkaline but in good balance. Some plants need an acid soil 4.5, and others thrive on an alkaline soil. But generally a soil condition reaction as nearly neutral as possible allows you to successfully grow the most plants.

In alkaline soils potash becomes less and less effective and eventually gets locked in. In very acid soils the element aluminum becomes so active that it becomes toxic to the plant. Acidity in soils controls many functions: it governs the availability of the food in the soil and determines the bacteria that thrive in it.

*Seed tapes are becoming popular with gardeners; in this method the seeds are already in the tape ready for planting.* (Photo courtesy George W. Park Seed Co.)

*Seedlings up and growing from seed tape installation.* (Photo courtesy George W. Park Seed Co.)

To lower the pH of soil (increase the acidity), apply ground sulfur at the rate of 1 pound/100 square feet. This lowers the pH symbol of loam soil about one point. Spread the sulfur on top of the soil and water it in.

To raise the pH of soil (sweeten it), add ground limestone at the rate of 10 pounds/150 square feet. Scatter the limestone on the soil or mix it well with the top few inches of soil and water. Add ground limestone or hydrated lime in several applications at 6- to 8-week intervals rather than using a lot at one time.

After preparing and testing the soil, rake the area with a *light* hand. A few lumps don't hurt anything and are better than a smooth terrain. For best results sow seeds in rows; if you just scatter them (broadcast), you will have a hard time telling the small plants from weeds when everything starts to sprout—by sowing in rows you will know at a glance which are your plants and which are nature's. For the distance between the rows follow the instructions on the package; it is important to have enough space between plants so seeds can grow and bloom in place.

Lightly mist the soil before sowing the seeds. As you sow, remember to label plants accordingly (use a *waterproof* pencil or marker). Fine dust-like seeds like petunias do not have to be covered, and large seeds need only a bare cover of soil. Remember that air and light must get to the seeds. Avoid soil caking too, or seedlings will have a difficult time emerging—keep the seed bed moist and friable.

Remove weeds as soon as they appear. They are easy to remove when small, but difficult when large. When seedlings are about 3 inches grown, thin them out so the strong ones have more growing space. *Don't* throw away the thinnings; these can be placed elsewhere or given to friends.

# 4. *Artificial Light*

Gardening by artificial light has attracted legions of followers in recent years, and it is an excellent way to start seeds and for growing seedlings.

There are many sources of artificial light—neon, mercury vapor, mercury fluorescent, quartz—but incandescent and fluorescent lights are generally used by gardeners to start seeds. These light sources are inexpensive, easily available, and, until more is known about light and plants, the most beneficial.

Generally a combination of both incandescent and fluorescent light is recommended for balanced lighting of plants. However, seeds can also be grown for considerable periods under either a fluorescent lamp or an incandescent bulb. The important thing to remember is that artificial light, whether incandescent or fluorescent or a combination of both, will stimulate and help plants grow.

## How Plants Use Light

A plant's duration of exposure to light—photo-periodism—determines the amount of food produced and whether the plant grows well. Some plants, like geraniums and orchids, need a great deal of light, but most room plants, like philodendrons and schefflera, tolerate low light levels and still flourish.

The visible spectrum has colors ranging from red to violet, like a rainbow. Plants require blue, red, and far-red rays to produce normal growth. Blue enables them to manufacture carbohydrates, and the red controls assimilation and affects the plant's response to the relative length of light and darkness. Far-red rays work in conjunction with red rays in several ways: they *control seed germination* and also control stem length and leaf size by nullifying or reversing the action of red rays.

Fluorescent lamps emit the rays of the blue part of the spectrum; incandescent lamps supply the needed far-red rays.

If you wish to grow only foliage plants, use fluorescent or incandescent light alone. African violets and other gesneriads will blossom under fluorescent light, but with perennial and annual seedlings the rate of growth will be greatly increased if incandescent light is added.

### Duration and Intensity of Light

The dark period (*absolute* darkness) is the crucial time for plants. Although we know what plants fall into categories of short-day, long-day, and day-neutral spans, we do not know where *all* the plants belong in the pattern. Gardenias, poinsettias, chrysanthemums, and Christmas begonias need 9 to 13 hours of light to bloom. Long-day plants like roses, carnations, some begonias, African violets, gloxinias, and geraniums require 12 to 18 hours of illumination. Room plants such as philodendrons, podocarpus, and pittisporums need 12 to 14 hours of light.

Light intensity is measured in foot-candles. Summer sunlight often exceeds 10,000 foot-candles, but on cloudy days summer light may only be 1,000 foot-candles. Many foliage plants grow well with very little light (about 150 foot-candles); with these plants bloom is not a prime factor as long as the plant continues to grow. But most flowering plants need from 500 to 1,000 or more foot-candles. A new light meter manufactured by the General Electric Company may be used for making foot-candle measurements. However, these meters cannot measure the effectiveness of Gro Lux or Plant Gro lamps, which are lacking in green and yellow output. If using these lamps you must watch the plants to determine if they are getting too much or not enough light.

### Light Setups

You can make your own fluorescent light setup, but there is an array available at suppliers. Individual units can be mounted over shelves or use boxes or prefabricated shelved carts and table models

*This artificial-light setup includes seed sowing, seedlings, and also mature plants.* (Photo courtesy George W. Park Seed Co.)

*Gro-Lux lamps are being used by many gardeners to start seeds and to help promote growth in seedlings. This unit, called the "Gro-Tower," is made by the manufacturer GTE Sylvania.* (Photo courtesy GTE Sylvania Inc.)

fitted with fluorescent light. Just what you use depends on how much space you have, but generally you will find a table-tray setup the best.

There are many types of fluorescent lamps (cool white, warm white, etc.) and lamps made especially for plant growth, such as Gro Lux and Plant Gro. I prefer the Gro Lux lamps, although friends have had equal success with seedlings by using warm white lamps.

No matter what lamps you use, there must be ample light. A single tube will do litle good; you need at least two fluorescent lamps for maximum growth.

For growing seeds under artificial light you can use the same seed pans, flats, or pans as for other indoor sowing. Some gardeners cover the seed container with plastic, but we find no protection is necessary, and humidity is just as good without the covers.

Once seeds are sown, place the trays 6 to 8 inches from the light. Be sure to keep the growing medium—perlite, vermiculite, etc.—moderately moist at all times. If you are using a "soilless" mix be sure to supply moderate feedings after true leaves appear.

Try to maintain a uniform temperature in the range of 65° to 70° F. Provide at least 14 to 16 hours of artificial light, and use a timer to regulate the light rather than relying on your own judgment. Keep a buoyant atmosphere, for a stagnant condition causes rot in seedlings.

When the second set of true leaves appears, transplant seedlings to individual pots or flats and then return them to their places under the lamps. Keep the tops of the plants at least 8 inches from the lamps; most commercial units provide adjustable devices for the fixtures.

*This commercial artificial-light table model shows mature plants. However, seeds and seedlings can also be grown under such a setup.* (Photo courtesy Floralite Co.)

### Germinating Seeds and Rooting Cuttings

Use 10 lamp-watts/square foot of growing area. Place the light source 6 to 8 inches above the propagating medium or root cuttings. A light period of 16 hours produces satisfactory results. A longer light period (up to 20 hours) has also been used with good results. Recent studies indicate that light-sensitive seeds exposed rather than lightly covered produce a higher germination percentage. You can also cover the seeds lightly with soil or vermiculite and then comb or scrape the surface lightly after soaking the cover medium. This allows better penetration of light as well as incorporating air into the seed areas.

When the seedlings are to be transplanted to the outdoor environment, take the proper steps to harden them. Gradually reduce the temperature and keep the plants somewhat drier than usual until they are acclimated to the outside conditions. A common procedure is to expose the seedlings outside during the day and to take them in for the night until they are firm enough to stay outside permanently. It may take about one and a half weeks to fully harden the seedlings. During the first few days, however, the young plants need some protection from the sunlight. Remember to provide plants with sufficient air circulation, especially if they are in protected areas. This will prevent the buildup of high temperatures and reduce the wilting of the young seedlings.

### Transplanting to Growing Bed

Use 20 lamp-watts/square foot of growing area. The photo-period depends on the light requirements of the individual plant and some experimenting will be necessary. Locate the fixtures so that light will fall upon each row of plants as well as between the rows (light between the rows provides some light energy to the lower leaves of the plants). Connect the fixtures to pulleys so you can raise and lower lights to adjust the intensity and to maintain a minimum amount of shading from sunlight. Follow accepted cultural and environmental practices for species and varieties of plants grown; control temperature, humidity, ventilation, and pH factor, and optimum levels of carbon dioxide enrichment, water, and plant food.

# 5. Cuttings

Getting new plants from cuttings is easier than most people think. The secret is to take the cutting at the proper time and then to give it good care.

A cutting is a small section of a plant treated to stimulate root growth. A cutting can be a portion of a stem, root, or leaf. Different species propagate in many different ways; these variables tend to confuse both beginning and experienced gardeners. With cuttings (in most cases) the new plant is *identical* to the parent plant. Shrubs and many evergreens are best propagated by cuttings; it's easier and faster. And many new plants can be started in a small space from just a few parent plants.

## Types of Cuttings

Depending upon the part of the plant from which they are taken, cuttings can be classified broadly as: root, hardwood, softwood, semihardwood, and leaf. Because some plants can be propagated by several different types of cuttings, the type you use depends on the easiest and least expensive means. Some plants cannot root from soft cuttings; they must be hardened first, when the first flush of soft growth is over.

Cuttings of some plants can be rooted in water, but most cuttings need a rooting medium such as coarse sterile sand, or a combination of sand and peat moss, or a combination of sand and vermiculite. The cutting can be started in a flower pot, a shallow box, or even a discarded aquarium. Make sure any container you use is 4 inches deep. To provide humidity, which most cuttings need, invert a jar

## STEM CUTTINGS

Take fresh, robust stems for cutting. ①
Trim bottom leaves.

② Cut bottom section of stem.

Dip into Root Hormone powder. ③

④ Set cuttings in flats of sterile potting medium; Water thoroughly.

over a flower pot (or put a Baggie in place). If you use shallow boxes (flats) or an aquarium, put a pane of glass 2 inches shorter than the actual size of the container over the top.

A hotbed or cold frame may also be used for cuttings during the mild seasons of the year if plants are shaded and protected from drafts.

There are many rooting mediums for cuttings, but a well-aerated sandy loam seems the most satisfactory. Be sure the soil mix is sterile. Plain sand can also be used with good results, as can vermiculite, peat, shredded sphagnum moss, or a combination of these materials. Use hormone- or growth-producing substances, powder or liquid (Rootone, etc.), for cuttings. These preparations can be found at suppliers.

*Root*

Many plants that have thick and fleshy roots can be started by root cuttings. In fall, take the thick end of the fleshy root attached to the plant and remove the lower portion. Cut the first end square and the last end at a 45° angle about 2 inches down. Place the cutting in soil in flats or in beds outside. Cover with half an inch of soil. It's a good idea to cover the cuttings with a light layer of leaves or hay (a mulch) to protect them from weather. When leaves appear in spring, the new plants may be taken and set into permanent places in the garden. Plants propagated by root cuttings include: windflower, butterfly weed, plumbago, bleeding-heart, summer phlox, poppy.

*Hardwood*

Take a hardwood cutting from a mature wood stem of the previous season's growth, branch, or twig; it is an easy and inexpensive way of vegetative propagation. No special equipment is needed. Take the cutting when the plant is dormant in winter or very early spring. Select firm, stiff, and unbendable growth. Use cuttings 4 to 12 inches long with at least two nodes; make the basal cut just below a node (the point where the leaf is attached to the stem) and the top cut about 1 inch above a node. (Three different kinds of cutting are used: mallet, heel, or straight cutting.) Deciduous ornamental shrubs

## HARDWOOD CUTTINGS

Take cuttings from dormant ①
one-year-old cane.

Treat cuttings with Rootone. ②

e cuttings and place horizontally ③
flat with sterile potting medium.

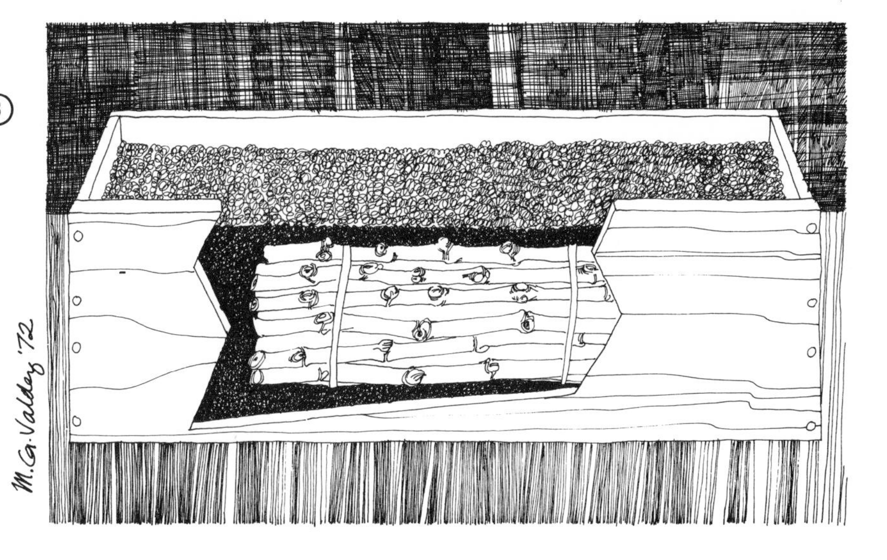

can be readily started from hardwood cuttings, e.g. forsythia, honeysuckle, and spirea; include at least two nodes on these cuttings.

There are various methods for handling hardwood cuttings, but for the home gardener this procedure seems best: tie the cuttings in bundles with wire or string, with the tops placed in one direction. Bury cuttings in sandy soil in a well-drained area or in sand in large boxes under cool and moist (40° to 50°F) conditions. Set the bundle in place horizontally. In spring, after calluses have formed at the end of the cuttings, put them in the ground vertically, with the callused end of each cutting several inches below the ground.

Another somewhat easier method is to take the cuttings during the dormant season, bundle and wrap them in sphagnum moss or peat (damp), and store them at 35° to 40° F. until spring, when they can be put in the ground. Do not let the cuttings dry out or become wet in storage. Check them occasionally; if you see buds developing, lower the storage temperatures. At planting time, cuttings with developed buds will form leaves before roots appear.

You can also take cuttings in fall, in which case you should plant them immediately. However, with this method winter freezes and thaws may cause heaving and plants will be harmed.

*Hardwood cuttings being placed in trench for propagation.* (USDA photo)

## SOFTWOOD CUTTINGS

ariety of softwood cuttings.

rowing medium being put in coldframe (LEFT). Pack down medium with brick or wood (CENTER). nsert cuttings (RIGHT).

*Softwood*

Take softwood cuttings (also called greenwood cuttings) from active growth of the current season before it hardens, generally in spring or early summer. Stems should be soft and succulent, never hard. Deciduous or evergreen species may be used; these include weigela, spirea, and forsythia, for example. Take softwood cuttings *with leaves attached.* Cuttings require ample humidity and warmth, 72° to 80° F. Try to select, as noted, a cutting that is flexible but strong enough to break when bent. Avoid weak or heavy shoots. Be sure the cutting is about 5 inches long with a least two nodes. Remove leaves on lower portion of cutting but retain the foliage on upper portion so it can manufacture food for the cutting while it roots. Place softwood cuttings in moistened rooting medium with two or three nodes below the ground. Get them planted as soon as possible. Even a few hours in sun may kill them. Keep the cuttings moist in flats or in a cold frame if you have one. Provide ample humidity. Shade until roots form and growth starts; then give them ample light. Transplant the cuttings as soon as they are well rooted and show growth.

Many herbaceous plants can be propagated by this method and are sometimes referred to as herbaceous cuttings, but the process is the same. Some plants started from softwood cuttings include: begonia, candytuft, chrysanthemum, impatiens, lantana, verbena.

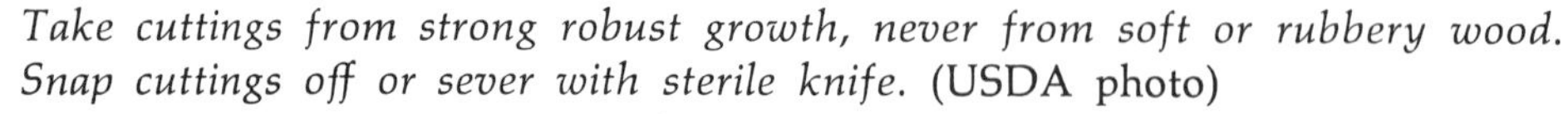

*Take cuttings from strong robust growth, never from soft or rubbery wood. Snap cuttings off or sever with sterile knife.* (USDA photo)

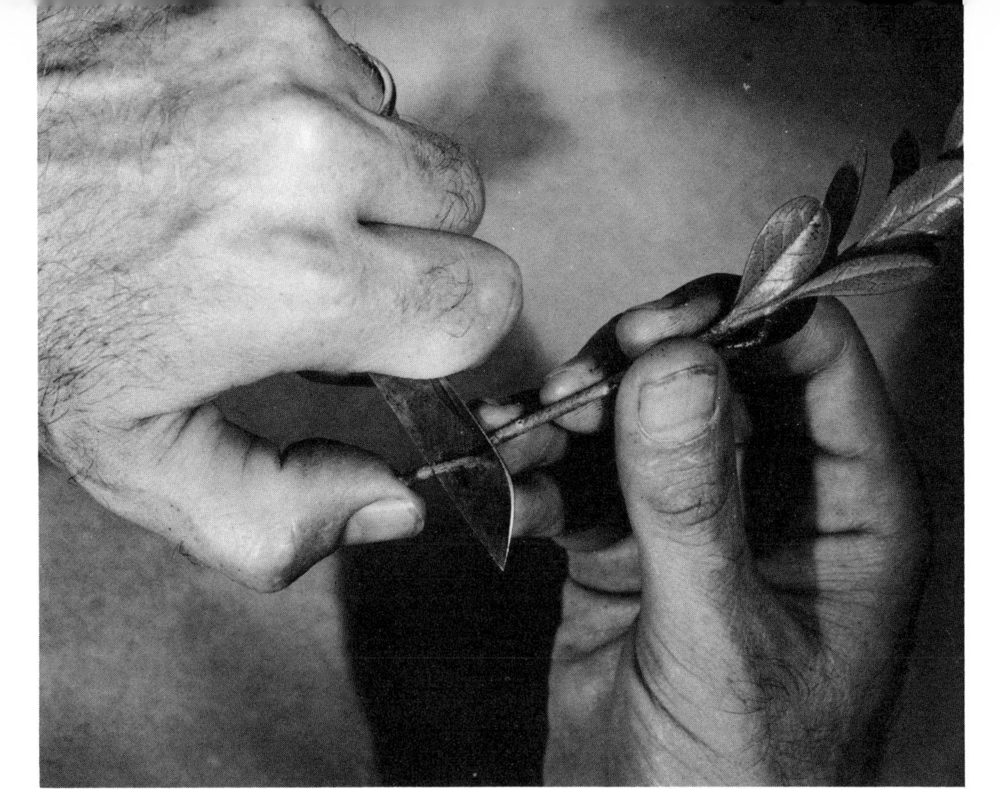

*With a sharp knife, take cuttings 4 to 6 inches long from the plant. Remove leaves from the lower third of the cutting.* (USDA photo)

*In a few weeks, you can gently pull up cuttings to see if roots have formed. If so, they are ready for planting. If not, reset cuttings and wait a few more weeks.* (USDA photo)

*Here's a way to make your own little greenhouse for cuttings in a plastic bag. Open occasionally so there is ventilation.* (USDA photo)

*Semihardwood*

Take semihardwood cuttings from broad-leaved evergreens or from partially mature wood of deciduous plants. Take cuttings in summer just after peak growth and when wood is almost completely matured. The best cuttings are pliable but brittle when bent. Make the cuttings 3 to 6 inches long and remove leaves from the lower end (but leave them on the upper end). House them in a box or cold frame where they have ample humidity and light, but not in sun. Provide constantly moist soil. As with softwood cuttings, once leaves show, the cuttings are ready for transplanting. Plants suitable for semihardwood cuttings include: abelia, camellia, cotoneaster, pyracantha, mahonia, pittosporum, holly, and evergreen azaleas.

*Leaf*

This propagation method is used frequently for house plants. Use the leaf blades with petioles (stem) of African violets, begonias, and many other popular house plants. Or partially insert pieces of leaves in sand.

## LEAF CUTTINGS

Take leaf cuttings from mature plants. ①

② Set leaves in sterile potting mix; weigh down with small pebbles so leaf is in contact with soil.

When small plants appear with roots, ③ cut them off for potting.

④ Set seedlings in containers of porous soil.

Fleshy-leaved plants such as begonias and violets should be handled in the following way: Cut a large vein on the under surface of the leaf with a sterile razor blade; make only a partial cut. Place the leaf flat against the propagating mix, with the natural upper surface of the leaf exposed. Hold down the leaves with small pebbles. Give cuttings shade and high humidity; in a short time new plantlets will form where each vein was cut. A variation of this method is to cut wedges of leaves (including a vein) and to insert them upright in sand. Plantlets will develop from the large vein at the base of the piece of leaf. Use whole leaves for African violets and insert them partially in sand. With enough moisture and humidity, new plants will form at the base of the petiole of the leaf blade. Remove plantlets and pot separately when leaves appear and roots have formed.

# 6. Layering, Grafting, and Other Propagating Methods

Layering is the development of roots on a stem that is still on the parent plant. Once the new plant starts growth, it is cut and grown on its own. There are various kinds of layering: simple, tip, mound, trench, and air. These are all quite simple procedures and provide an ideal way for the home gardener to get new plants with no cost and little effort—all sorts of shrubs and trees can be increased by layering.

The other propagation processes are by division, grafting, or rooting runners, stolons, offsets, or suckers.

## *TYPES OF LAYERING*

### *Simple*

Bend a branch of a mature plant to the ground and cover it partially with rooting or soil medium. Notch or make a small slit in the section to go underground. This will help to stimulate root development. Always leave the terminal end exposed. Be sure the branch stays horizontal to the ground—hold it in place with U-shaped vine clips. Cover the branch with 3 to 6 inches of soil.

Layering is best done in spring with dormant one-year-old shoots. After the growing season in fall, or the following spring before growth starts, the new plant should be adequately rooted and ready to be detached. Handle it then in the same manner as a rooted cutting of the same plant. Plants you can increase this way include: azalea, candytuft, cotoneaster, forsythia, rhododendron, yew.

*The process of simple layering is shown here.* (USDA photo)

*To air layer a plant, make a notch first.* (Photo by J. Barnich)

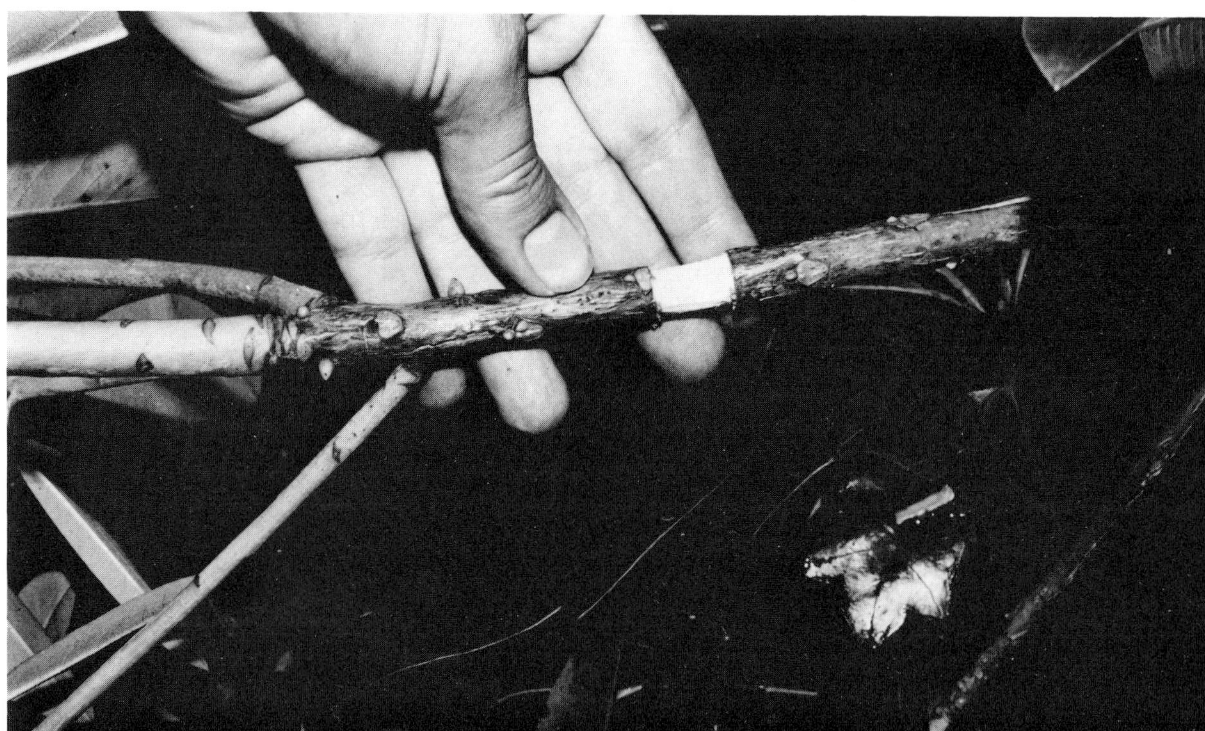

*Apply a ball of moist sphagnum to the notch.* (Photo by J. Barnich)

*Wrap the layered section in plastic and tie with string.* (Photo by J. Barnich)

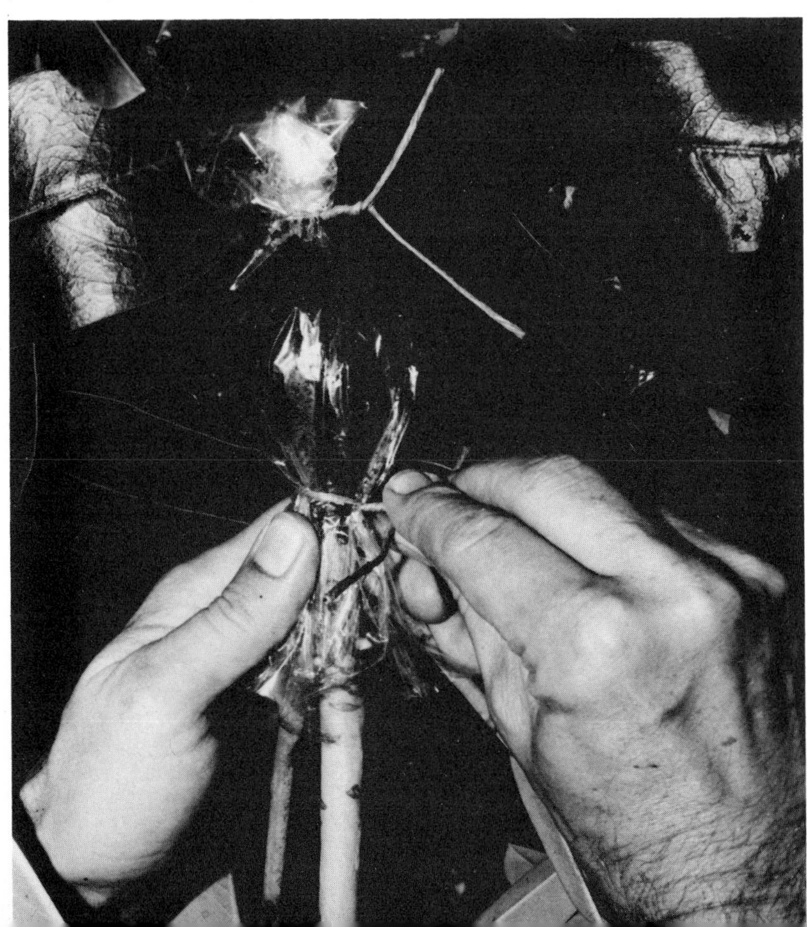

# TIP LAYERING

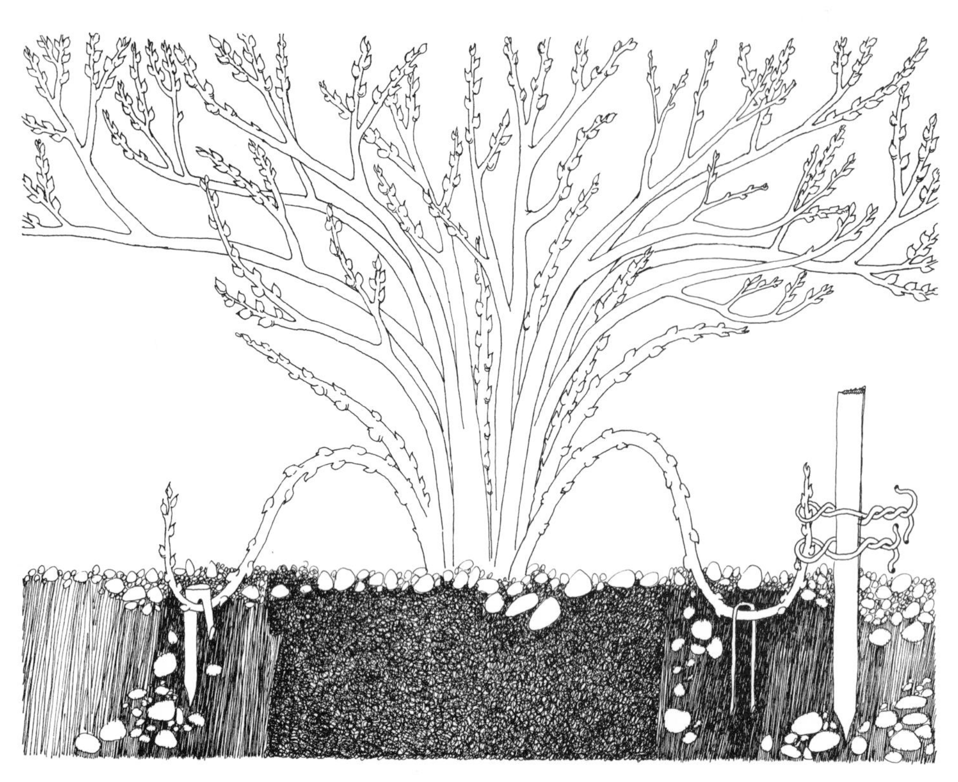

① **Bend shoots into ground.**

② **Cover with soil and set in place with wire.**

③ **When roots appear remo young plant and pot in porous soil.**

*Tip*

With tip layering the current season's shoot is bent to the ground. It grows into the ground but then recurves to grow vertically. To tip layer, make a hole 3 or 4 inches deep and insert the end of the shoot; then cover with soil. Generally, rooting takes place fast, and the new plants are ready for digging the end of the growing season.

*Mound*

Cut a plant back to the ground in the dormant season and put soil around the base so newly developed shoots can form roots and eventually new plants. This type of layering should be done with plants that are not flexible and which also have many shoots from the crown year after year. Such plants include currants, gooseberries, and some apple stocks.

*Trench*

Trench layering is a variation of simple layering. It is a way to produce many plants from one parent plant, and although seldom used by home gardeners, deserves mention in this day of high prices.

Grow a branch of a mature plant in a horizontal position in a trench filled with soil; you will have to bend the branch into position and anchor it down with wooden pegs or U-shaped wire devices. Add soil at intervals to cover the base of developing shoots. At the end of the growing season and when plants are dormant, remove the soil from the layered shoots. Then cut the plantlets from the original stock (as close to the base as possible) and treat as new plants.

*Finished air layering; when roots form, new plant is ready to be planted.* (Photo by J. Barnich)

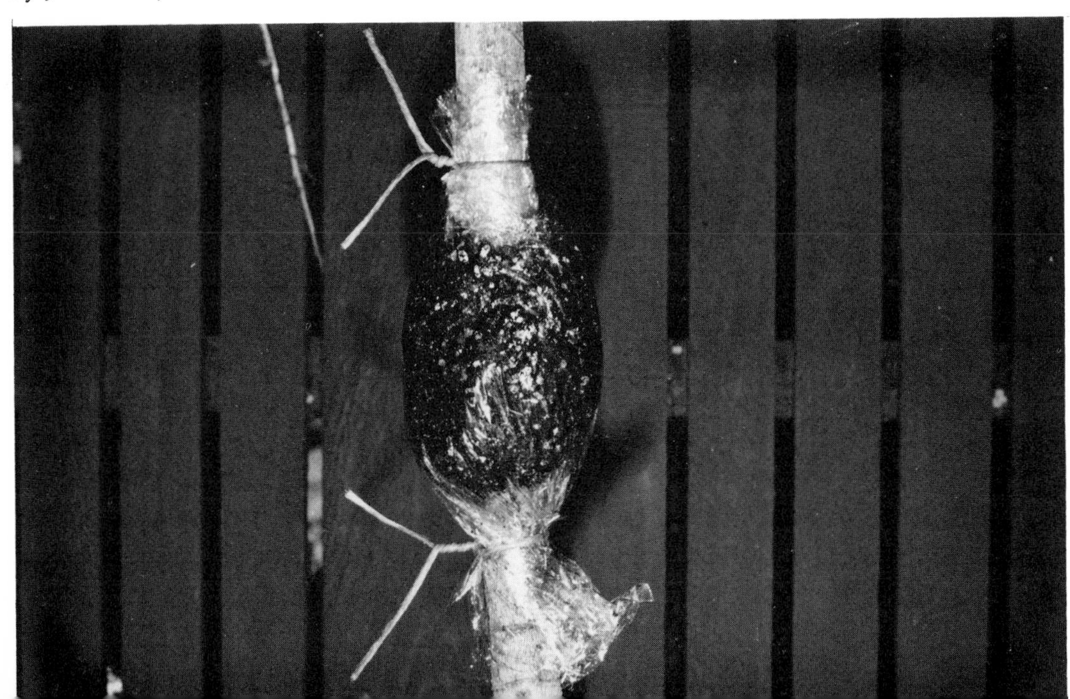

## AIR LAYERING

① Notch one-inch band around stem. Do not cut through stem. Discard bark.

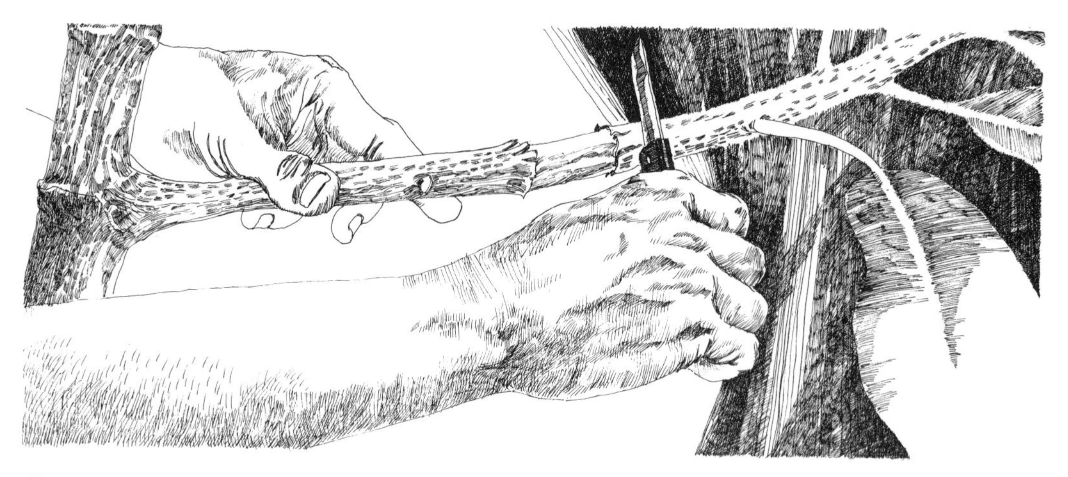

② Cover entire notch with ball of osmuda.

③ Cover osmuda with plastic sheet and tie both ends.

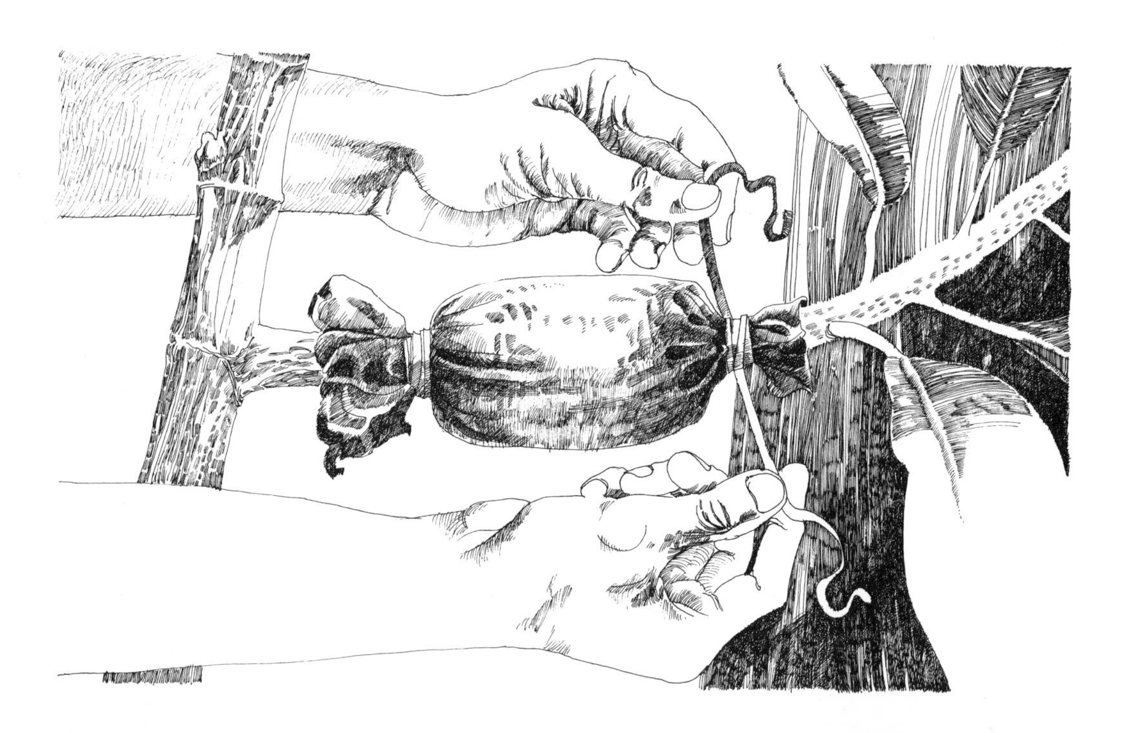

*Air*

Air layering (developed from the ancient Chinese version of *marcottage*) is really a very simple process if you want only a few plants of a given variety. It is slow and cumbersome for large-scale operations but fine for the home gardener. Choose a well-developed shoot and slice and girdle a 1- to 3-inch section; a complete girdle is best when cut down to the cambium (the thin tissue between the bark and wood) layer. Remove the section of bark and treat the wound with a hormone powder. Wrap the girdled area in a ball of sphagnum moss and cover it with a sheet of polyethylene plastic. Bind the plastic with string or rubber bands at both top and bottom. The rootball should be fully covered, *not* exposed, so moisture, along with high humidity, is retained in the moss, which encourages rooting. In 6 to 9 months roots can be seen in the moss ball; at that point the new plants may be severed and removed.

Air layering is used to propagate many subtropical and tropical trees and shrubs and is generally 100 percent successful. It is an easy way to ensure new house plants from such old ones as rubber plants, philodendrons, and so forth.

Do air layering in the spring on wood of previous season's growth or in late summer with partially hardened shoots. Pot the rooted layer in a suitable container in cool and humid conditions.

## Grafting

Grafting is still another part of plant propagation; it is joining together parts of two different plants so they unite and continue their growth as one plant. Its principal use is to perpetuate varieties that are not successfully reproduced by cuttings, division, or layering. It is an important part of fruit- and nut-tree propagation, but since it is used more by nurserymen than home gardeners, we shall only briefly discuss the subject.

The process of grafting involves putting a piece of the desired plant (a scion) and the understock (parent plant) in unison. The idea is to fit two pieces of living plant tissue together in such a way that they will grow and subsequently become one plant. Cover all cut surfaces with tape or hot grafting wax (from suppliers) to prevent drying out. Tend the graft until it has grown together. Then remove the top of the understock—the final result is a graft.

## POTTING OFF-SETS OR STOLONS

① Sansevieria ready for stolon separation.

② Pry apart root stock; Cut stolons with knife.

Pot-up individually in fresh soil; Water.

To graft plants successfully, choose plants closely related; non-related plants, such as a peach and a maple, cannot be grafted. The cambium must be carefully fitted to the stock (the lower portion of the graft, which developes the root system of the grafted plant). To ensure success, do the grafting procedure at the proper time of year, generally when growth starts in early spring.

Methods of grafting include whip and tongue, side, and cleft. Other grafting techniques are possible too, but for our purposes these three are sufficient.

### Runners, Offsets, Suckers

Runners are prevalent in many house plants, like the strawberry plant (Saxifraga) and chlorophytum. A runner is a stem that develops at the crown of a plant and grows horizontally from this area. New plants form at the nodes on the runner and take root while still

*This palm is being propagated by removing a stolon.* (Photo by M. Barr)

*At the base of the mother plant, this baby* Pandanus Veitchi *has grown almost as large as the mother plant. It is ready to be severed and potted separately. (Photo by author)*

attached to the mother plant. It is a simple procedure to take the runners when they have roots and plant them separately.

Offsets or stolons are branches that develop from the base of the parent plant. They generally look like thickened stems with rosettes of leaves; offsets are prevalent in the bromeliad family. Once the offset is developed (with leaves), cut it off close to the main stem. It is then ready to be potted like any rooted cutting.

Suckers on a plant come from below the soil in the vicinity of the crown. Some plants produce suckers freely, but others produce none. When the suckers have formed roots and leaves, dig and cut them from the parent plant. Do not pull them off; cut them cleanly with a sterile knife. Treat the new plant as a cutting.

*Division is a simple method of propagation; merely pull the clump apart (but gently) and plant.* (Photo by J. Barnich)

*Iris rhizomes being severed for new plants.* (Photo by J. Barnich)

*Here a small tulip bulb is being taken from the mother bulb.* (Photo by J. Barnich)

### Division

Plants that form crowns can be easily propagated by just dividing the crown, one of the simplest propagation methods. Some shrubs and a great many herbaceous perennials respond well to this vegetative method. Dig plants and cut them into sections or gently pull them apart by hand and plant separately. Division is employed frequently with perennials and also with such shrubs as jasmine and English boxwood.

## DIVIDING RHIZOMES

① Cut back leaves to one-third.

② Divide rhizome with sharp knife.

③ Cover roots with soil, firm; Water.

## DIVISION

① Clumps of mature plants can be divided to more new plants.

Separate by cutting or gently ② pulling crowns apart; be sure to include roots with each crown.

③ Put new plants in porous soil mixture; water copiously.

# 7. *The Plants*

Gardens are expensive; even small properties require a good number of plants to make them attractive. The advantage of growing your own is not only one of economics, as noted earlier, but also and perhaps more important, there is great satisfaction in knowing that the plants in your garden are your very own, grown and nurtured by you to perfection.

## Trees and Shrubs

Surprisingly, trees and shrubs grow faster than you think; in only a few years most shrubs will be ready for pruning, and your seedling trees will be making a showing. And make no mistake about it, virtually every garden needs a good framework of trees and shrubs to make it attractive.

If you have grown your own perennials and annuals, you certainly should try some trees and shrubs. The process of propagating these plants is not much more difficult once you know something about the seeds and their requirements. Seeds of some trees and shrubs are hard-coated and often require scarifying or soaking to hasten germination. Other seeds must be sown as soon as they are ripe, and so on.

In the following list are requirements for various trees and shrub seeds and some hints to help you grow them successfully. Also included are various propagating methods for the plants.

*Trees and shrubs are necessary for beauty and grace; they are the framework of a good landscape.* (Theodore Brickman, LA)

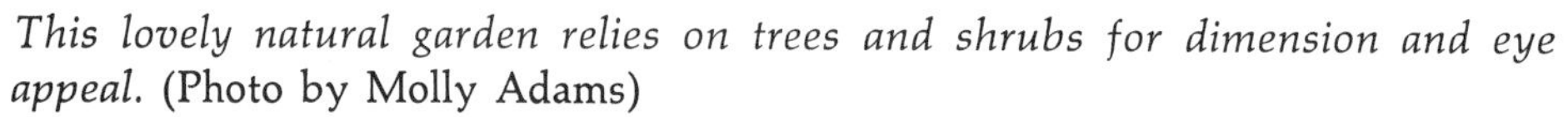

*This lovely natural garden relies on trees and shrubs for dimension and eye appeal.* (Photo by Molly Adams)

## TABLE OF TREES

| BOTANICAL AND COMMON NAME | PROPAGATION METHOD | HINTS |
|---|---|---|
| *Abies balsamea** (balsam fir) | Seed | Use fresh seed or stratify for 60–90 days |
| *Acacia baileyana** (Bailey acacia) | Seed | Soak in boiling water for 20 hours |
| Acer (Japanese maple) | Seed | Stratify 60–120 days; sow in spring |
| | Hardwood cuttings | Some species respond |
| *Aesculus carnea* (red horse-chestnut) | Seed | Stratify 120 days |
| | Layering | Low-growing types respond |
| *Ailanthus altissima* (tree of heaven) | Root cuttings | Choose female trees only; male trees smell bad |
| | Suckers | Take readily in spring |
| *Albizzia julibrissin* (silk tree) | Seed | Sow soon as ripe; soak in boiling water for 24 hours |
| | Root cuttings | Plant in early spring |
| *Alnus incana* (speckled alder) | Seed | Sow in fall or spring |
| | Cuttings | From mature wood |
| *Betula papyrifera* (canoe birch) | Seed | Sow in fall or in spring after stratifying 4-8 weeks |
| | Leaf cuttings | In summer |
| *Catalpa speciosa* (western catalpa) | Seed | Germinate readily; sow in late spring |
| | Softwood cuttings | In summer |
| *Cedrus atlantica** (atlas cedar) | Seed | Soak in water |
| | Hardwood cuttings | Take in late summer or fall |
| *Celtis occidentalis* (hackberry) | Seed | Sow in fall or stratify 60–90 days for spring sowing |
| | Cuttings | Works sometimes |
| *Cercis canadensis* (eastern redbud) | Seed | Stratify 60 days |
| | Softwood cuttings | In spring or summer |
| *Chamaecyparis obtusa** (Hinoki false cypress) | Seed | Stratify 60 days |
| | Cuttings | Sometimes respond |
| *Chionanthus virginica* | Seed | Stratify 90–120 days |

| | | |
|---|---|---|
| *Cornus florida* (flowering dogwood) | Ripe wood cuttings | Take in fall or spring |
| *Cryptomeria japonica** | Seed | Sow soon as ripe; germinates in 4 months |
| | Cuttings | In sand in fall; takes 1 year to root |
| *Elaeagnus angustifolia* (Russian olive) | Seed | Stratify 90 days and sow in spring |
| | Root cuttings | Sometimes responds |
| | Layering | Often responds |
| *Eucalyptus globulus** (blue gum) | Seed | In spring as soon as ripe |
| *Fagus sylvatica* (European beech) | Seed | Stratify 90 days |
| *Fraxinus americana* (white ash) | Seed | Stratify 60–90 days |
| *Gingko biloba* (maidenhair tree) | Softwood cuttings | In spring |
| *Gleditsia tiacanthos* (sweet locust) | Seed | Scald in water; sow in spring |
| | Hardwood cuttings | In spring |
| *Koelreuteria paniculata* (goldenrain tree) | Softwood cuttings | In spring |
| | Root cuttings | In spring |
| *Laburnum watereri* | Seed | In spring; germinates easily |
| *Liquidambar styraciflua* (sweet gum) | Seed | Stratify 30–60 days in summer |
| | Softwood cuttings | In summer |
| *Liriodendron tulipifera* (tulip tree) | Seed | Stratify 60 days |
| *Magnolia soulangeana* (saucer magnolia) | Seed | Stratify 120–180 days |
| *Malus baccata* (Siberian crab apple) | Seed | Stratify 30–90 days |
| *Phellodendron amurense* (cork tree) | Seed | Germinates readily; fall |
| | Root cuttings | In spring |
| *Picea abies (excelsa)** (Norway spruce) | Seed | Stratify 60–90 days |
| *Pinus densiflora** (Japanese red pine) | Seed | Stratify 30–90 days |
| *Platanus acerifolia* (London plane tree) | Seed | Let seed overwinter on tree; collect in early spring and sow |

| | | |
|---|---|---|
| *Populus alba* (white poplar) | Seed | Plant at once |
| | Hardwood cuttings | Some root readily; plant in spring |
| | Softwood cuttings | Some root when planted in summer |
| *Quercus alba* (white oak) | Seed | White-oak seed ready to germinate as soon as mature in fall; black oak needs stratification |
| *Robinia pseudoacacia* (black acacia) | Seed | Soak 20–120 minutes in hot water before sowing |
| | Root cuttings | Occasionally successful with some varieties |
| *Salix alba* (white willow) | Root cuttings | Easy |
| | Stem cuttings | Easy |
| *Sorbus aucuparia* (European mountain ash) | Seed | Stratify 60–120 days |
| *Taxus baccata** (English yew) | Cuttings | Easy (seeds too difficult) |
| *Thuja occidentalis** (American arborvitae) | Cuttings | Take in summer |
| *Tilia americana* (American linden; basswood) | Seed | Stratify for 90–120 days; difficult |
| *Tsuga canadensis** (hemlock) | Seed | Stratify 60–90 days or plant in fall for spring germination |
| *Ulmus americana* (American elm) | Softwood cuttings | In spring |

*Evergreen trees.
For stratification process see Chapter 2.

## TABLE OF SHRUBS

| BOTANICAL AND COMMON NAME | PROPAGATION METHOD | HINTS |
|---|---|---|
| *Abelia grandiflora* (glossy abelia) | Leafy cuttings | Take from matured growth |
| | Hardwood cuttings | In spring or fall |
| *Amelanchier canadensis* (shadblow service berry) | Seed | Stratify 90–180 days |
| *Andromeda polifolia* (bog rosemary) | Seed | Germinates easily |

*Some popular shrubs.*
(Photos by author)

*Rhododendron*

*Azalea*

*Camellia*

| | | |
|---|---|---|
| *Arbutus unedo* (strawberry tree) | Seed<br>Cuttings | In early spring<br>In autumn |
| *Arctostaphylos uva-ursi* (bearberry) | Cuttings | In spring |
| *Berberis koreana* (Korean barberry) | Softwood cuttings | In spring |
| *Buddleia davidii* (summer lilac) | Softwood cuttings | In spring |
| *Buxus sempervirens* (common boxwood) | Softwood cuttings | In spring or fall |
| *Callistemon citrinus* (bottlebrush) | Leaf cuttings | Seedlings rarely satisfactory |
| *Calluna vulgaris* (heather) | Seed<br>Leaf cuttings | In spring<br>Any time of year |
| Camellia | Seed, cuttings, layering | Cutting easy to root |
| *Ceanothus americanus* (New Jersey tea) | Seed, cuttings, layering | Soak in boiling water |
| Cotoneaster | Seed<br>Leaf cuttings | Difficult<br>In spring |
| *Daphne odora* (fragrant daphne) | Leaf cuttings<br>Layering | Take in late spring<br>In spring |
| Deutzia | Hardwood cuttings<br>Softwood cuttings | Either method easy |
| Euonymus | Leaf cuttings | Take mature wood |
| *Forsythia intermedia* (border forsythia) | Hardwood and softwood cuttings | Easiest from hardwood cuttings |
| *Gardenia jasminoides* (Cape jasmine) | Leaf cuttings | Take from fall to spring |
| *Hamamelis vernalis* (spring witch hazel) | Seed | In spring |
| *Hibiscus syriacus* (shrub althea) | Hardwood cuttings | Take in fall; store until spring |
| *Hydrangea arborescens* | Hardwood and softwood cuttings | Take in early spring |
| *Ilex cornuta* (Chinese holly) | Softwood cuttings | Spring, fall |
| *Jasminum officinale* (poets' jasmine) | Hardwood cuttings | Easy |
| *Kalmia latifolia* (mountain laurel) | Seed<br>Softwood cuttings | In spring<br>Works well with most |

| | | |
|---|---|---|
| *Kerria japonica* | Hardwood cuttings | Generally easy |
| *Lagerstroemia indica* (crape myrtle) | Hardwood cuttings<br>Seed | In spring<br>Occasionally successful |
| *Laurus nobilis* (sweet bay) | Seed | Sow immediately |
| *Ligustrum amurense* (Amur privet) | Hardwood cuttings | Take in spring |
| *Lonicera fragrantissima* (winter honeysuckle) | Hardwood cuttings | Take in spring |
| *Mahonia aquifolium* (holly mahonia) | Seed<br>Leaf cuttings | Stratify through winter<br>Generally successful |
| *Nerium oleander* (oleander) | Seed<br>Cuttings | Plant immediately<br>Trim mature wood |
| Philadelphus | Hardwood cuttings<br>Softwood cuttings | In spring<br>In summer |
| *Photinia serrulata* (Chinese photinia) | Seed | Stratify 30-60 days |
| *Potentilla fruticosa* (cinquefoil) | Seed, hardwood cuttings | Take in autumn |
| *Pyracantha coccinea* (scarlet firethorn) | Softwood cuttings | Spring, fall |
| Rhododendron | Seed, cuttings, grafting, layering | Cuttings difficult; grafting generally successful |
| Rosa | Softwood and hardwood cuttings, layering | All asexual methods generally easy |
| *Spiraea prunifolia* (bridal wreath spiraea) | Hardwood cuttings | Use root-promoting substances |
| *Syringa vulgaris* (common lilac) | Hardwood cuttings | Difficult |
| *Viburnum davidii* | Seed | Complicated seed dormancy conditions |
| | Cuttings | Cuttings generally successful |
| | Grafting<br>Layering | |
| *Weigela florida* | Hardwood cuttings<br>Softwood cuttings | In spring<br>From spring to fall |
| Wisteria | Softwood cuttings | In midsummer |

*Some popular perennials.*
(Photos by author)

*Kniphofia*
*(torch lily)*

*Gazanias*

*Lobelia*

*Another popular perennial.*
(Photo by author)

*Dianthus*

Perennials

Perennials, the mainstay color guard of the garden, are plants that die to the ground after frost and sprout again in spring. Many perennials are long living and grow for many, many years, but some, e.g. lupines and delphiniums, last only 3 or 4 years.

There are several ways of propagating perennials: from seed; from leaf cuttings, division, or layering; or from stem or root cuttings. However, most modern varieties of perennials are hybrids and will not come true from seed, so other propagation methods are favored.

Cold frames are generally used to sow seed; but of course flats and other containers based on the same construction principles can be used. Press very fine seed into the soil (cover larger seed with soil to a depth of two to three times the diameter of the seed). After sowing, mist the soil, and shade plants from direct sun. Once seeds germinate, let them have plenty of air and light; after the first true leaves are up, thin them out. Most perennials' seeds will not bloom until the second season.

You can also propagate perennials from stem cuttings, but division, as explained above, is much easier and faster for the home gardener. Layering (Chapter 6) can also be used for perennials.

To keep perennials healthy, divide them every few years; replant only the strong growth from the outer edge of the clump. Generally, divide fall-flowering plants in the spring and spring-flowering plants in the fall.

The following perennials are the ones we have worked with, and we offer them as suggestions for your garden. Included are notes on seed germination and also best vegetative methods. The table is not complete, and no doubt we have missed some of your favorites, but this is a matter of space, not of preference.

## *TABLE OF PERENNIALS*

| BOTANICAL AND COMMON NAME | OPTIMUM TEMPERATURE FOR SEED GERMINATION | WEEKS RE-QUIRED FOR SEEDS TO GERMINATE | VEGETATIVE METHODS |
|---|---|---|---|
| *Achillea ptarmica* (yarrow) | 68°–70° F. | 1–2 | Division in spring or fall |
| *Allium pulchellum* | 68°–74° F. | 3–4 | Division |
| *Alyssum saxatile* (golden tuft) | 68°–86° F. | 3–4 | Division in spring or fall |
| *Althaea rosea* (hollyhock) | 68°–70° F. | 2–3 | Division |
| Anchusa (bugloss) | 68°–86° F. | 3–4 | Division in spring or fall; root cuttings |
| *Anemone coronaria* (poppy anemone) | 68°–70° F. | 5–6 | Division |
| *A. pulsatilla* (pasqueflower) | 68°–70° F. | 5–6 | Division |
| Aquilegia (columbine) | 68°–86° F. | 3–4 | Division in spring |
| Arabis (rock-cress) | 60°–68° F. | 1–2 | |
| Armeria (thrift) | 68°–70° F. | 3–4 | Division |
| Artemisia (dusty miller) | 50°–55° F. | 1–2 | Division |
| *Asclepias tuberosa* (butterfly milkweed) | 68°–86° F. | 3–4 | Division |

| | | | |
|---|---|---|---|
| Aster (michaelmas daisy) | 68°–70° F. | 2–3 | Division in fall or spring |
| *Aubrieta deltoidea* | 50°–55° F. | 2–3 | Stem cuttings |
| Begonia (tuberous) | 68°–72° F. | 2–3 | Leaf, stem cuttings; division of roots |
| *Bellis perennis* (English daisy) | 68° F. | 1–2 | Division |
| *Campanula carpatica* (Carpathian harebell) | 68°–86° F. | 2–3 | Division in spring or fall |
| Canna | | | Division of rhizome |
| *Cerastium tomentosum* (snow-in-summer) | 68°–70° F. | 2–4 | Division |
| *Cheiranthus cheiri* (wallflower) | 54°–56° F. | 2–3 | |
| Chrysanthemum | 68°–70° F. | 2–4 | Division; softwood cuttings |
| Coreopsis | 68°–70° F. | 2–3 | |
| *Cyclamen indicum* | 68°–70° F. | 3–4 | Division |
| *Delphinium grandiflorum* (delphinium) | 54°–56° F. | 3–4 | Division |
| Dianthus (pinks) | 68°–74° F. | 2–3 | Layering, division |
| *Dicentra spectabilis* (bleeding heart) | * | 6 | Division; stem cuttings; root cuttings |
| *Dictamnus albus* (gasplant) | * | 6 | |
| Digitalis (foxglove) | 68°–86° F. | 2–3 | Division in spring |
| Echinops (globe thistle) | 68°–86° F. | 1–4 | Division; root cuttings |
| Erigeron (midsummer aster) | 68°–70° F. | 3–4 | |
| *Gerbera jamesonii* (transvaal daisy) | 68° F. | 2–3 | Division |
| Geum (avens) | 68°–86° F. | 3–4 | Division |
| *Helenium autumnale* (sneezeweed) | 68° F. | 1–2 | Division |
| *Helianthemum nummularium* (sunrose) | 68°–86° F. | 2–3 | Division; summer cuttings |

| | | | |
|---|---|---|---|
| Heliopsis | 68°–86° F. | 1–2 | Division |
| Helleborus (Christmas rose) | * | | Division |
| *Hesperis matronalis* (sweet rocket) | 68°–86° F. | 3–4 | |
| *Heuchera sanguinea* (coral bells) | 68°–86° F. | 2–3 | Division; leaf cuttings |
| *Lathyrus latifolius* (pen vine) | 68°–86° F. | 2–3 | |
| *Lavandula officinalis* (lavender) | 52°–90° F. | 2–3 | |
| Linum (flap) | 54° F. | 3–4 | Division |
| *Lobelia cardinalis* | 68°–86° F. | 3–4 | Division |
| Myosotis (forget-me-not) | 68° F. | 2–3 | Division; cuttings |
| Nierembergia (cupflower) | 68°–86° F. | 2–3 | Division; cuttings in full |
| Oenothera (evening primrose) | 68°–86° F. | 1–3 | Division |
| *Ornithogalum thyrsoides* (Star of Bethlehem) | 68°–78° F. | 2–6 | Division |
| *Papaver nudicaule* (Iceland poppy) | 54° F. | 1–2 | |
| *P. orientale* (oriental poppy) | 54° F. | 1–2 | Root cuttings in late summer |
| Pelargonium (geraniums) | 68°–76° F. | 2–8 | Cuttings |
| Penstemon (beard tongue) | 68°–86° F. | Slow, uneven | Division |
| *Phlox divaricata* | * | | Division; root cuttings |
| *P. paniculata* (garden phlox) | * | | Division; root cuttings |
| *Platycodon grandiflorum* (balloon flower) | 68°–86° F. | 2–3 | Division in spring |
| Primula (primrose) | 68°–74° F. | 3–6 | Division; cuttings in spring |
| Pyrethrum (painted daisy) | 68°–76° F. | 2–3 | Division |

| | | | |
|---|---|---|---|
| Ranunculus | 68°–72° F. | 1–4 | Division |
| Rudbeckia (coneflower) | 69°–86° F. | 2–3 | Division |
| *Sinningia speciosa* (gloxinia) | 68°–72° F. | | Leaf cuttings |
| Tigridia (tiger flower) | 68°–74° F. | 2–4 | Division |
| *Trollius europaeus* (globeflower) | * | | Division |
| *Verbena canadensis* (clump verbena) | 54°–90° F. | 2–4 | Cuttings |
| Veronica (speedwell) | 54°–90° F. | 2–3 | Division |
| *Vinca minor* (periwinkle) | 68° F. | 2–3 | Cuttings |
| *Viola cornuta* (violet) | 54°–90° F. | 2–3 | Division; runners; cuttings in spring |

*Needs stratification.

### Annuals

Generally, annuals are plants that complete their entire life cycle in one growing season and then die. They are known for their color and minimum care—once you have your seedlings planted it's only a matter of plenty of water and lots of sunlight to bring them to perfection.

Start annuals indoors to get a head start on spring (as outlined in Chapter 2), or sow them directly in the garden when weather permits. Because some annuals like petunias and snapdragons need a long growing season they cannot complete their life cycle in northern parts of the country. Thus sowing seeds indoors or in cold frames is a logical answer. Other annuals, such as marigolds and zinnias, should be sown directly in the garden in almost all parts of the country.

Germination times for annuals vary; some start growth quickly, but others may take several weeks. Once again, very fine seed should not be fully covered; it should be sprinkled thinly on the growing medium. Larger seeds require a light dusting of soil.

When to plant annuals depends on what part of the country you live in (most seed packets furnish this information). Whether they thrive or just grow will depend on such cultural factors as soil, sunlight, moisture, and, in general, climate itself. New varieties of annuals are introduced frequently—seed is available through many suppliers who advertize in garden magazines.

The following table of annuals is by no means complete, but there are enough to get you started in the right direction.

## *TABLE OF ANNUALS*

| BOTANICAL AND COMMON NAME | OPTIMUM TEMPERATURE FOR SEED GERMINATION | WEEKS REQUIRED FOR SEEDS TO GERMINATE | HINTS |
|---|---|---|---|
| Agathea (felicia) (blue daisy) | 68°–70° F. | 2–3 | Start indoors |
| *Ageratum houstonianum* | 68°–86° F. | 3 | Seedlings fragile |
| *Alyssum maritimum* (alyssum) | 68° F. | 2–3 | Start indoors |
| Amaranthus | 68°–86° F. | 3–4 | Plant directly in garden |
| *Anchusa capensis* (forget-me-not) | 68°–70° F. | 1–2 | Self-sows |
| *Antirrhinum majus* (snapdragon) | 60°–65° F. | 1–2 | Pinch plants |
| *Arctotis stoechadifolia* (African daisy) | 68°–70° F. | 2–3 | Plant directly in garden |
| *Begonia semperflorens* (wax begonia) | 68°–86° F. | 1–3 | Fine bedding plant |
| Browallia | 70°–75° F. | 2–3 | Hanging plant |
| *Calendula officinalis* (pot marigold) | 68°–70° F. | 2–3 | Likes coolness |
| *Callistephus chinensis* (China aster) | 68° F. | 2–3 | Sow indoors |
| *Celosia argentea cristata* (cockscomb) | 68°–86° F. | 1–2 | Resents transplanting |
| Centaurea (dusty miller) | 68°–86° F. | 3–4 | Start indoors |
| Chrysanthemum (annual) | 68°–70° F. | 2–4 | Good cool-region plant |

*Some popular annuals.*

*Petunias*
(Photo courtesy Pan American Seed)

*Zinnias*
(Photo courtesy Burpee Seed Co.)

*Marigolds*
(Photo courtesy Burpee Seed Co.)

| | | | |
|---|---|---|---|
| *Clarkia elegans* | 54°–70° F. | 1–2 | Good cool-region plant |
| *Cleome spinosa* (spiderflower) | 54°–90° F. | 1–2 | F1 hybrids excellent |
| *Cosmos bipinnatus* (cosmos) | 68°–86° F. | 1–2 | Easy annual |
| Cynoglossum (Chinese forget-me-not) | 68°–80° F. | 1–2 | Lovely annual |
| *Delphinium ajacis* (larkspur) | 60°–65° F. | 2–3 | Plant directly in garden |
| Dianthus (pink, carnation) | 68°–70° F. | 2–3 | Easy; start indoors |
| Dimorphotheca (cape marigold) | 68°–70° F. | 2–3 | Cool-region plant |
| Eschscholzia (California poppy) | 54°–70° F. | 2–3 | Plant directly in garden |
| *Euphorbia variegata* (snow-on-mountain) | 68°–74° F. | 1–2 | Handsome |
| Gaillardia (blanketflower) | 68°–72° F. | 2–3 | Easy; start in garden |
| Gazania | 68°–72° F. | 2–3 | Robust |
| Godetia | 68°–70° F. | 2–3 | Takes some shade |
| Gypsophila (baby's breath) | 68°–70° F. | 2–3 | Best started in garden |
| *Helianthus annuus* (sunflower) | 68°–86° F. | 2–3 | Grows quickly; start outdoors |
| Helichrysum (strawflower) | 68°–78° F. | 1–2 | Needs long season; start indoors |
| Heliotropium (heliotrope) | 68°–86° F. | 3–4 | Start in garden |
| *Impatiens balsamina* (snapsweet) | 68°–70° F. | 2–4 | Many new varieties |
| *I. Sultana* (impatiens) | 68°–70° F. | 2–3 | Start indoors |
| *Ipomoea purpurea* (morning glory) | 68°–86° F. | 1–3 | Crack seed coat |
| Linaria (toadflap) | 54°–60° F. | 2–3 | Easy |
| *Lobelia erinus* | 68°–86° F. | 2–3 | Difficult to transplant |

| | | | |
|---|---|---|---|
| *Mathiola incana* (stock) | 54°–90° F. | 2 | Sow indoors |
| Matricaria (feverfew) | 74°–86° F. | 1–2 | Easy |
| Mirabilis (four-o'-clock) | 68°–86° F. | 1–2 | Easy |
| *Molucella laevis* (bells of Ireland) | 86°–90° F. | 2–3 | Self-sows |
| Nemesia | 60°–74° F. | 2–3 | Good edge plant |
| Nicotiana (flowering tobacco) | 76°–80° F. | 2–3 | Do not cover seed |
| Nigella (love-in-a-mist) | 86°–90° F. | 2–3 | Plant directly in garden |
| Papaver (Shirley poppy) | 55° F. | 1–2 | Short bloom season |
| Petunia | 68°–70° F. (Some varieties need higher temperatures) | 2–3 | Bountiful flowers |
| *Phlox drummondii* (annual phlox) | 58°–60° F. | 2–3 | Needs coolness for germination |
| *Portulaca grandiflora* (moss rose) | 68°–86° F. | 2–3 | Needs warmth for germination |
| *Reseda odorata* (mignonette) | 54° F. | 2–3 | Not easy |
| *Salpiglossis sinuata* (painted tongue) | 68°–86° F. | 2 | Needs warmth for germination |
| *Salvia splendens* | 68°–86° F. | 2–4 | Good red color |
| Scabiosa (pincushion clower) | 68°–86° F. | 2–3 | Needs warmth for germination |
| Schizanthus (butterfly flower) | 54° F. | 1–2 | Start in ground |
| Tagetes (marigold) | 68°–86° F. | 1–2 | Self-sows |
| Thunbergia (clockvine) | 68°–86° F. | 2–3 | Start indoors |
| Thymophylla (Dahlborg daisy) | 68°–84° F. | 2–3 | Popular |
| Tithonia (Mexican sunflower) | 68°–86° F. | 2–3 | Start in garden |

| | | | |
|---|---|---|---|
| *Tropaeolum majus* (nasturtium) | 68° F. | 1–3 | Easy; start in garden |
| *Verbena hortensis* (verbena) | 68°–86° F. | 3–4 | Start indoors; needs long growing season |
| Viola (pansy) | 60°–74° F. | 1–2 | Self-sows |
| *Zinnia elegans* (zinnia) | 68°–86° F. | 1–2 | Start in garden |

### Bulbs

Bulbs are buried treasure because their beauty goes on year after year. Too many people believe that once a bulb finishes blooming, that is the end of it—but this is far from the truth. Bulbs last for years and are readily increased by natural separation. Mature bulbs produce a number of bulbils or offsets. These small bulbils can be removed and planted when the mother bulb is dug up. Thus, once an initial investment is made in, say, lilies or narcissus, you can have flowers for years to come. You can also grow bulb plants from seed, but it usually takes several years for plants to mature.

Bulbs also include tubers (e.g. tuberous begonias and dahlias) and corms (e.g. gladiolus and achimenes). There are, of course, differences in each, but basically they are entire-blooming plants in neat natural packages called bulbs. For the most part bulbous plants can live off their own storehouse of food for some time, but they should never be neglected when in active growth.

Corms are shortened rhizomes, that is, thickened bases of stems generally underground. Each year, new corms are formed on top of the mature corm and can be planted separately. A corm differs from a bulb in that the greater part of a bulb is not stem but scales, really thickened bases of leaves. Tubers are thickened portions of stems beneath the soil. Tuberous roots differ from tubers in that they have no eyes from which growth can start.

Mail-order suppliers list all plants with fleshy counterparts under the term "bulbs." This catch-all category includes bulbs, tubers, and corms. But no matter how you define these plants, the makings of the flowers are already in the bulbs, and all you do is to plant them.

Some bulbs are winter hardy (need cold weather to grow) and thus should be left in the ground year after year. Other bulbs must be planted and lifted each year. Spring-blooming bulbs are the most

*Some popular bulbous plants.*

*Anemone*

*Anemone*

*Ranunculus*

popular because they are easy to manage and bear color in early spring.

Generally, in most parts of the United States summer-flowering bulbs must be dug up in fall and stored over winter. They can be left in the ground only if the temperature does not go below freezing.

*Comments*

Here are general propagating instructions on the more popular bulbous-type plants:

*Gladiolus* (gladiola). Popular garden plants because they seem to grow in any soil without too much attention, but they do need sun. In fall, dig up and store the corms if you are in a cold climate; in mild climates they can remain in the ground. When the foliage turns brown, cut off the tops, and then dry the bulbs in an airy, shady place for a few weeks. You will see new corms on top of the old dried withered ones; cut and store the new corms in paper sacks in a dark place at about 45° to 50°F. until planting time.

*Lilium* (lily). Lilies are synonymous with grace and beauty, and in recent years hybridists have given us some stellar plants. To increase your lily supply, lift the plants and divide them; do this about 4 weeks after they bloom. Replant the bulbs after a reasonable length of time. Exposure naturally weakens them.

There are several ways to increase lilies: the easiest is just to divide the established bulbs (as mentioned). Or use the scale method, that is, remove four or five outside thick scales of old bulbs at replanting time. Put the scales in trenches about 4 inches deep and cover them with sand.

*Tulipa* (tulip). If you think tulips are anything like they were even a few years ago, you are in for a surprise. There are many new varieties, and the tulip has climbed from an ordinary flower to an extraordinary one. There are parrot kinds, fringed ones, tall ones, small ones, cottage-type, Darwin, Mendel, and so on. Some are so utterly beautiful they stagger the imagination.

Plant bulbs in a prepared soil bed between October and the middle of November. They can be left in the ground or lifted for other plantings if you desire. In that case, lift them with roots and leaves, and then keep them in a shady place to ripen. When the foliage has turned brown, store them in a cool, dry place until planting time in the fall. Tulips need cold weather; in all-year temperature climates, you must buy pre-cooled bulbs or store bulbs in the refrigerator for 6 weeks before planting them.

*Dahlias* (dahlia). These plants have tuberous roots and rest between seasons. The next year's flower is produced by the fleshy extension of the old stem. For best results, grow dahlias in a sandy soil; give them plenty of water and sun. Feed them regularly and put some bone meal in the soil. As soon as the tops are killed by the first frosts in the fall, cut the plants back to about 4 inches above the crown; dig them up in a few days. Dry them in a well-ventilated place for a day before storing in peat moss or vermiculite in a cool, but not freezing, area. In spring, divide the bulbs, allowing one eye or bud to each root, and replant.

*Begonia tuberhybrida* (tuberous begonia). For sheer drama and intense summer color, these plants are star performers. Tuberous begonias today have been bred to near perfection in flower form, size, and color. The choice is vast, one prettier than the other; Camelliaeflora, Ruffled, Cristata, Fimbriata, and Marginata in two forms (*Crispa marginata*, with frilled single flowers, and *Double marginata*, with petals lined and edged with a band of contrasting color). Other forms include Narcissiflora, Picotee, and Rose.

After blooming, when leaves turn dry and yellow, water plants sparingly; let growth continue for as long as possible. When foliage is completely yellow, dig up tubers, wash off soil, remove stems, and place them in an airy, sunny place for a few days. Then store them in a cool, frost-free location until it is time to divide and start them again from division of tubers.

# 8. House Plants

It would be difficult to find a home without at least a few house plants, whether they are African violets or philodendrons. Today's indoor plants are indeed treasures, especially for apartment dwellers, and I myself would not be without them.

Because the rising cost of house plants and specimen plants, more than ever it is economical and wise to start increasing your plants. You may argue that you do not have the space nor time to fool with these projects. My answer is "You do!" Even the smallest plant can be increased in 5 minutes. And beyond the thought of money there is great joy and fascination in growing your own plants for guests to see. For those people who think a great deal of botanical knowledge is needed to propagate plants, I repeat what I mentioned in Chapter 1: "Even a child can do it."

### Means of Propagation

The methods of increasing house plants are generally the same as for outdoor plants, namely, leaf cuttings, division, air layering, etc. (We rarely propagate indoor plants from seed. We like to see results in a relatively short time; in some cases, you must wait years with seeds.)

The easiest way to increase plants is to take a cutting and insert it in clear water in a mayonnaise jar and place it on a bright but not sunny window sill. When it shows root, pot it. Rhizomatous begonias, syngoniums, and some philodendrons can be easily increased by this method. Others, such as dieffenbachias and tradescantias, can be propagated by taking the stem cuttings (good sturdy shoots) and placing them in vermiculite or sand in a warm humid place. Breadboxes, baking dishes, casseroles—any container with a cover—make good breeding cases, as do clay pots with jars or plastic Baggies over them. Take 4-inch pieces just below a leaf node (the little swelling on the stem). Remove the leaves from the lower 2 inches. Dip the cutting in hormone powder to stimulate growth, and insert one-third to one-half its length in the growing mix. Keep cuttings warm, from 75° to 80°F. Lift the cover for an hour or so to ventilate plants.

Plants such as African violets, kalanchoes, and sedums can be increased by leaf cuttings. This is really a simple and fascinating way to get more plants. Select healthy leaves, and make small cuts across the veins (use a sterile knife or razor blade). Then put the leaf on sterile vermiculite. Be sure the leaf is in contact with the growing medium; we weigh the leaf down with small stones. Put the cuttings in a warm and bright, but not sunny, place. Keep them covered part of the day with plastic. Plantlets appear from the slit veins and draw nourishment from the mother leaf. When the plants are showing two sets of leaves, put them in 2- or 3-inch individual pots, using a rich and friable house-plant soil.

Large foliage plants like philodendrons, ficus, and dieffenbachias are the mainstay of the indoor gardener's garden. Mature plants are expensive; there is no reason why you cannot start a few youngsters yourself. Take 4-inch cuttings of mature stems. Coat both ends with sulfur, lay the pieces in the rooting medium, cover lightly with

Rhoeo spathacea *is a favorite house plant; offsets or stolons are produced by mature plants and can be potted separately.* (Photo courtesy Merry Gardens)

# STEM CUTTING

① Cut at base.

Insert in sand. ②

③ Cover with plastic bag.

*Dieffenbachia is another popular house plant; this plant could be divided to make two specimens.*

*A fine indoor subject is the bromeliad* Bromelia balanse; *note the many baby plants it produces. This large specimen is summered outdoors.* (Photo by author)

the mix, and firm them in place. New plants form from dormant eyes along the canes. Remember to furnish adequate humidity (60 percent) and warmth. When plantlets are large enough to handle, put them in individual pots of friable soil. (Air layering as described in Chapter 6 can also be used to propagate these larger plants.)

Many plants such as chlorophytum, episcia, and saxifraga, which are all good window plants by the way, produce runners. Take runners about 3 inches long and handle them like cuttings. Roots develop and plants are ready for individual pots of soil in about 4 weeks to 4 months, depending on the plant. Offsets or stolons are somewhat similar to runners. These little plants develop at the base of the mother plant. When they are 2 or 3 inches high they may be removed and potted separately. Some plants that produce these babies are bromeliads, gesneriads, and agaves.

With lovely indoor ferns, propagation is usually done by division. However, many people enjoy growing ferns from spores. Spores look like small dark spots on fern fronds (and are often mistaken for insects). But spores, like seeds, carry the germ of life for new ferns. Fern leaves, however, do not come directly from spores. Spores first develop little green heart-shaped plants called *prothallia*; later, fern leaves are produced.

*Chlorophytums produce runners that can be taken from the mother plant and potted separately.* (Photo courtesy Merry Gardens)

To get your own ferns from spores, select mature leaves and lay them with spore surface down on a piece of white paper. Cover the fronds with a newspaper or blotter to absorb moisture. In a few days, the spores will naturally separate from the frond. Spores can then be sown in flats or similar containers with 3 inches of soil; dust the spores on the surface of the soil and then cover the container with plastic to provide adequate humidity. Keep soil moist, but never wet. Remove plastic for a few hours a day if condensation occurs. When true fern leaves have sprouted, tiny plants can then be put in their individual pots.

The following house plants are only a small selection of the many kinds suitable for propagating.

*This mature plant of Schefflera could easily be divided to make new plants.* (Photo courtesy Potted Plant Information Center)

## *TABLE OF HOUSE PLANTS*

| BOTANICAL AND COMMON NAME | VEGETATIVE METHODS | SEEDS | HINTS |
|---|---|---|---|
| Abutilon (flowering maple) | Cuttings | • | Needs ample humidity |
| *Acalypha hispida* (chenille plant) | Cuttings | | Needs good humidity |
| Achimenes | Tubers | • | Requires plenty of heat |
| Adiantum (maidenhair fern) | Division; spores | | Easy with division or spores |
| Aeschynanthus (lipstick-vine) | Cuttings (layering) | | Subject to damping-off fungus |
| *Agave americana* (century plant) | Offshoots | | More plants than you will need |
| Aglaonema (Chinese-evergreen) | Cuttings (air-layering) | • | Cuttings easier to do than layering |
| Alternanthera | Cuttings; division | | Difficult to start |
| Aphelandra | Cuttings | | Plants get leggy so take cuttings yearly |
| *Araucaria excelsa* (Norfolk Island pine) | Stem cuttings (air layering) | • | Layering works well, but allow roots to form |
| Asparagus (asparagus fern) | Division | • | Simple with division |
| Aspidistra (cast-iron plant) | Division | | Another easy one |
| *Asplenium nidus* (bird-nest fern) | Spores | | Best from spores |
| Begonia | Leaf cuttings | • | Most cuttings need ample humidity |
| *Beloperone guttata* (shrimp plant) | Tip cuttings | | Not always successful |
| Caladium | Tubers | | Start each year |
| Calathea | Division | | Watch for damping-off |
| Caryota (fishtail palm) | Offsets; division | | Slow but sure with either method |
| Ceropegia (rotary vine) | Cuttings | | Difficult |

| | | | |
|---|---|---|---|
| Chlorophytum (spider-plant) | Division or runners | | Put runners in water; simple |
| Cissus (vitis) | Cuttings | | Easy |
| Codiaeum (croton) | Cuttings | • | Need heat |
| *Coffea arabica* (Arabian coffee) | Cuttings | | Somewhat difficult |
| Coleus (painted-leaf-plant) | Tip cuttings | • | Needs ample humidity |
| Columnea | Tip cuttings | | Not always successful |
| Davallia (rabbit's-foot fern) | Division of rhizomes | | Easy |
| Dieffenbachia (dumbcane) | Cuttings (air layering) | | Air layering best |
| Dizygotheca (false aralia) | Cuttings (air layering) | | Difficult to start |
| Epiphyllum (orchid cactus) | Cuttings | | Subject to damping-off |
| Episcia (peacock plant) | Cuttings; division | | Needs ample humidity |
| *Euphorbia pulcherrima* (poinsettia) | Cuttings | | Difficult |
| *E. splendens* (crown-of-thorns) | Cuttings | | Easy |
| Ficus (fig or rubber plant) | Leaf cuttings (air layering) | | Air layering generally successful |
| *Gynura aurantiaca* (velvet plant) | Cuttings | | Temperamental |
| Hedera (ivy) | Cuttings | • | Easy with plenty of moisture |
| *Hibiscus rosa-sinensis* (rose-of-China) | Cuttings | • | Needs warmth and ample humidity |
| *Hoya carnosa* (wax plant) | Cuttings | | Difficult but not impossible |
| Jacobinia (king's crown) | Tip cuttings | • | Easy |
| Kohleria | Tip cuttings; division of rhizomes | | Needs good humidity |

| | | | |
|---|---|---|---|
| Maranta (calathea) (prayer plant) | Division; leaf-stalk cuttings | | Subject to damping-off |
| Medinilla (love plant) | Cuttings | | Difficult |
| Monstera (Swiss-cheese plant) | Cuttings (air layering) | | Air layering generally successful |
| *Nephrolepis bostoniensis* (Boston fern) | Runners; division | | Division easiest method |
| Pandanus (screw-pine) | Offsets; division | | Either method very simple |
| Pelargonium (geranium) | Cuttings | | Dust ends of cutting with hormone powder |
| Pellaea (cliff-brake or button fern) | Spores | | Easy in protected container |
| Peperomia | Stem or leaf cuttings | | Easy |
| Philodendron | Cuttings | | For most types just place in water |
| Pilea (aluminum plant) | Cuttings | | Cuttings usually successful |
| Polypodium (polypody) | Division | | Easy |
| Rebutia (crown cactus) | | • | Seed germinates easily under lights |
| Rechsteineria | Cuttings; tubers | • | Dividing tubers easiest method |
| Rhodea | Cuttings | | Will root in water |
| *Saintpaulia ionantha* (African violet) | Leaf cuttings; division | • | Simple |
| Sansevieria (snake plant) | Division; cuttings | | Easy with division |
| *Saxifraga sarmentosa* (strawberry-geranium) | Runners | | Place in water to root |
| Schefflera (Queensland umbrella plant) | Cuttings; half-ripened stems | • | Somewhat difficult |
| Scindapsus (pothos) (ivy-arum) | Cuttings | | Very easy |
| Smithiantha (temple-bells) | Offshoots; stolons | | Simple |

| | | | |
|---|---|---|---|
| Streptocarpus (cape-primrose) | Cuttings; division | • | Very easy |
| Syngonium (arrowhead) | Cuttings | | Grows like weed |
| *Vallota speciosa* (Scarborough-lily) | Offsets | | Pot offsets when they have two pairs of leaves |
| Veltheimia | Division of bulbs | | Repot every year |
| Zebrina | Root cuttings | | Will root in water |

# 9. Vegetables and Herbs

Today everyone is growing vegetables; we hope this fad becomes a habit because home-grown vegetables are one of the pleasures of life. They are really simple to cultivate, provided you select the kinds that are appropriate for your area. And you can grow enough, *more* than enough, in even a small plot of say 5 x 10 feet.

You may not be able to grow all vegetables—asparagus and corn are still beyond me in California's climate—but you can have carrots, beets, lettuce, onions, turnips, and spinach; all lend themselves to small sites. Early-maturing crops, which include lettuce, radishes, and turnips, will allow a second planting of other vegetables for fall.

Crops can also be grown vertically to conserve space; tomatoes, pole beans, and cucumbers can be grown this way. In fact, tomatoes and cucumbers can even be grown in pots if you are a city gardener. Thus there is no excuse for not having fresh vegetables. The initial procedures of planting the seed takes but a few minutes and cultivating the plants not much longer.

At this time we have a small vegetable garden, only 10 x 12 feet, and a variety of vegetables are thriving. The main requirements of these plants are good rich soil, at least 6 hours of sun daily, and frequent deep watering. Insects are sometimes a problem, yet we do not spray indiscriminately or at all if we can help it. Usually we bring in ladybugs every year (they eat scores of aphids), plant nasturtiums around the garden to repel insects, and encourage birds. Even though birds do some damage, their good (they eat grubs and insects) far outweighs their bad qualities.

Remember that a good many of the packages of seed you buy are already pre-treated with preventatives. Seed treating by the commercial grower protects even seeds planted under adverse conditions and allows them to grow and thrive.

### Planting and Care

After you have prepared the soil, rake it finely into a seed bed. Make shallow trenches or "drills" (furrows) of suitable depth, and keep them straight so that plants will be easy to cultivate. Distribute seeds evenly in the drills and scatter fine soil over them; then firm the soil. Note that certain vegetables like pole beans, cucumbers, and squashes must be planted in hills instead of drills—a hill is a circle of seeds about 12 inches in diameter.

Thin out young plants when they are large enough to handle. This gives the remaning plants room to develop. Do not neglect the thinning process; it is the only way to get a good yield. Once the garden is started, maintain weeding and cultivation to conserve moisture.

#### Tips

1. Make sure all vegetables (small, medium, and tall) are placed so they have sun.
2. To save space, place quick-growing crops between slow-growing ones.

*The author's vegetable garden; all plants were started from seed.* (Photo by author)

3. Rotate crops yearly to conserve nutrients in the soil.
4. Mulch with organic materials to control weeds and to conserve moisture.
5. Thin out the seedlings.
6. Use insect-preventing chemicals that don't leave a poisonous residue.
7. Harvest vegetables right before maturity; they will be at their prime when you eat them.

SOME FAVORITE VEGETABLES

Beans are available in endless varieties, and all require warm weather and warm soil. Plant lima beans (the ones we have been successful with here) 1 to 2 inches below the soil; they will germinate in about 120 days.

Beets, generally a disease-resistant crop, grow lavishly for me. Be sure soil is about neutral (pH 7). Cover seeds with a half inch of soil and plant 2 inches apart. When growth starts, be sure to thin plants, but don't throw them away: they make fine cooked greens. This is a cool-weather crop and germinates in 30 to 50 days.

Broccoli is a cold-weather crop and germinates in 30 days at the most. Plant seeds with a half inch of soil covering them as soon as soil conditions permit. Thin occasionally. Eliminate cabbage worms, which may attack crops, with rotenone dust.

There are endless varieties of cabbage available. Plant seeds about ¼ to ½ inch deep as soon as weather permits (cabbage does best in cool weather). Thin to stand about 16 inches apart. In optimum conditions germination of seed takes place in about a week.

Carrots, which *should* be an easy vegetable to grow, are occasionally tricky because they will not respond in poor soil; they need good friable soil. Sow seeds and cover ever so lightly; germination usually takes 2 to 3 weeks. Thin to stand 4 inches apart.

Cucumbers, a hot-weather favorite, bear in about 70 days from seeds. Plant seeds pointed end down and bury them about 1 inch deep. Germination occurs in about a week, and then thinning is necessary. Many varieties are available.

Lettuce is a favorite home crop and really does well in even poor conditions. Most are cool-weather crops, although a few have been developed lately for warm weather. Plant seed a half inch deep; you can expect lettuce in 20 to 40 days.

*Some popular vegetables.*
(Photos courtesy Burpee Seed Co.)

*Jersey King Eggplant*

*Burpees M&M Cucumber*

*Burbank Hybrid Sweet Corn*

Radishes are popular because they mature in about 25 days. Sow seed a half inch deep for germination in about 7 days. Plant in early spring and then again in fall if you desire. Many varieties are available.

Tomatoes are everyone's favorite. These plants like warm weather and a good rich warm soil to prosper. You can buy plants already started at nurseries, or sow seeds if you want particular varieties. Germination occurs in about a week. Set plants 2 feet apart, and stake them, or they become a mess.

Cauliflower is difficult and takes great care to get it going and to nurture it to maturity. Corn too does not do well for us here because our summers are not warm enough. Furthermore, corn takes more care than we have time for.

The following table summarizes planting conditions for favorite vegetables:

| | WHEN TO | | SPACING | | DAYS TO |
|---|---|---|---|---|---|
| PLANT | PLANT | DEPTH | PLANT | ROW | MATURE |
| Pole bean | After last frost | 2″ | 3′ | 4′ | 60 |
| Carrot | After last frost | 1/2″ | 2–4″ | 18″ | 100 |
| Celery | After last frost* | 1/4″ | 8″ | 30″ | 190 |
| Cucumber | After last frost | 1″ | 3–4′ | 3–4′ | 140 |
| Lettuce | After last frost | 1/4″ | 1′ | 18″ | 70–90 |
| Onion | After last frost | 1/2″ | 1–3″ | 1′ | 50–60 |
| Pepper | After last frost* | 1/2″ | 30″ | 3′ | 60–80 |
| Tomato | After last frost* | 1/2″ | 30″ | 3′ | 60–90 |

* Start indoors.

### Herbs

Harvesting herbs yourself is a satisfying experience and herbs are more flavorful and tastier than the ones you buy in stores. These attractive plants are easy to grow in average conditions, take little space, and can be grown directly in the ground or in pots. If you want herbs for flavor, keep them near the kitchen so you can use them fresh in stews or salads. (Some herbs, for example, lavender and verbena, are delightful just for their fragrance.)

Start seeds in a fairly sandy soil that drains readily. Place herbs so they will get at least 3 hours of sun. Keep the soil moist. Divide and prune the plants occasionally so they don't crowd out other plants.

Cut herbs just as the flowers are about to open; this is when the oils are the most abundant. You must properly cure and store herbs if you want to save them. Wash the leaves or stems in cold water and dry them thoroughly by spreading them over a wire mesh in a warm place. You can also place them on a baking sheet in a 200° oven; leave the door open. When the leaves are dry, strip them from the stems and put them in airtight containers. Be careful that you don't overdry the leaves; they should be barely crumbly. (A third method involves tying the stems in bunches and hanging them from the ceiling in a dark place until you want to use them.)

# 10. Hybridization

This final chapter could very well have started the book. Plant breeding is creating new plants. Generally, such creations are left to the professional growers, who spend years and much money to give the public plants that are robust or have large flowers, scent, color, or some outstanding characteristics. One or all of these factors is the prime reason for breeding plants. Whether grown by the professional or the amateur, the basic concept of hybridization remains the same: to create a better plant.

There is great fascination in playing nature, and although plant breeding may not appeal to everyone, it does intrigue countless gardeners. Plant breeding is an easy procedure; you do not actually have to know about genetics or chromosomes, although this knowledge is valuable. All you really need is a love of plants, an understanding of their cultural needs, and the desire to perhaps find a new plant. Remember that hybridization is the geneticist's livelihood; with you it is an enjoyable hobby.

### Methods

Hybridization is the transfer of pollen from one flower to another. The crosses are performed with the same kind of species of flowers. The beginner should stay within the same family of plants for his plant breeding. Plants from totally different families, e.g. a primrose or a dandelion, simply won't work. Start with a definite goal in mind—scent, larger flower, longer life of flower, or color—and stick to it; don't go off on a tangent. Hybridization takes patience and time.

*Hybridization is responsible for these beautiful coleus variations.* (Photo courtesy George W. Park Seed Co.)

To pollinate or cross a flower, take pollen grains and set them on the surface of the stigma, which is often sticky and receptive, so grains can adhere to it. Pollen is a spore; when it is placed on the top of the pistil it germinates and sends out a hair-like tube that goes into the pistil and to the ovary chamber below. It then discharges male cells to do the actual fertilizing (within the ovary is the embryo cell).

The mechanics of crossing are thus quite simple; it is the *timing* that is complex. The pistil and the pollen must be ripe for success. The pollen must be bursting from the anther, and the pistil must be fully formed and slightly sticky. Touch the first to the second—the cross is made. Touch a camel's hair brush or a pipe cleaner to two or three ripe anthers and gather the pollen; then brush the yellow dust on the pistil. [Once again, let me mention that you have to time the operation and gather the pollen when it is ripe and brush it on the pistil when it is ready (sticky).] Often pollen will be ready a few weeks before the pistil is. In that case, keep the pollen in a small dry bottle tightly capped; put a little calcium chloride in the bottle and then a layer of cotton to place the ripe anthers on. Cork the bottle and put it in the refrigerator. There is no way to keep a pistil in waiting.

Once the crossing operation is performed, be sure that no other pollen reaches that particular pistil. A bee's visit could ruin the entire cross, as can self-pollination. To keep insects away, tie a transparent bag over the flower before it is fully opened. To avoid a flower from self-pollinating, cut away the flower to expose the stamens; the stamens' pollen may complete with those you are putting in place. Leave a naked pistil rising from the base of the flower and then bag it.

*Most of the dieffenbachias shown here are products of the hybridizer's art.* (Photo by Albert & Merkel)

### Genes

As noted above, the mechanics of hybridization are relatively simple. However, selecting the right parents is something else again; it's vitally important to know the parentage of the plant you are working with and the parents' parents as well. Which characteristics are dominant and which are recessive? You want to get the best from the best by crossing the plants until the ultimate goal is attained. To accomplish this you must know something about genes.

Genes are living organisms that determine the size, growth, color and form, structure, number of flowers, etc. There are thousands of male and female genes in a plant, generally in pairs. It is these genes that determine the inheritance when you cross two plants. A single gene or a number of genes may control more than one or only one characteristic at a time. These genes remain independent even when crossed. Thus, a so-called recessive gene may assert itself in a later generation.

Gregor Mendel, the great pioneer in the hybridizing field, is called the father of genetics because he discovered a great many laws of heredity. One of his laws is the law of independent assortment, which explains what happens when the original parents differ in two or more distinct respects. By crossing plants (animals too), Mendel discovered that there were fixed laws of heredity characteristics. For example, when dwarf peas were crossed with tall ones, the offspring generation was *not* a variety of peas somewhere halfway between the two; the cross of the first generation reproduced all tall varieties with no trace of dwarfness. Since the dwarf characteristic did not come through, it was established as recessive, and the tall, because it did come through, as dominant. However, in further studies Mendel discovered that in the second generation the recessive characteristic reasserted itself to 25 percent. Thus, the second generation (known as F2) had 25 percent dwarfs and continued to pass on this dwarf trait to future generations. The other 75 percent were tall, but in future generations they broke up, and only one-third of the offspring was tall and retained that trait for succeeding years. The other two-third broke up again, and so on.

There is thus endless intrigue in hybridization, but you must keep intensive records and notes as you experiment, and put in a great deal of time. The chief aim of hybridization is to bring diverse, outstanding qualities together in one variety; this can be generally done by successive crossing rather than by one lucky attempt.

### Hints

As mentioned, stay within one plant family when you are hybridizing, and decide beforehand what your goal is. Would you like to see a blue orchid, a red daffodil, or a scarlet chrysanthemum? There are possibilities for many kinds of plants; further hybridization can bring about new and exciting varieties. Asters, chrysanthemums, irises, roses, and dianthus are all families within which there is room for improvement. Just what you decide on is a personal matter. The possibilities are extensive, but before you approach any particular plant family know all about it—its flowers and its history—and you will have a better chance of being successful.

Various plant societies publish yearbooks with information on varieties and present breeding work. There are many fine books on plant breeding; some are listed in the Bibliography. Horticultural libraries offer additional references.

# Glossary

AIR LAYERING A method of *propagation.* A twig or *shoot* attached to the parent plant is wrapped in moist sphagnum moss or polyethelene plastic so that it will form roots and can later be removed and replanted.

ANNUAL A plant whose life span is only one season.

ANTHER The organ borne at the upper end of the *stamen.* It develops, secretes, and discharges pollen.

ASEXUAL PROPAGATION Reproduction from vegetative parts of a plant.

AXIL The point of divergence between a branch or leaf and the axis from which it arises.

BONE MEAL Crushed and ground bones used as plant fertilizer.

BROADCAST To sow (scatter) seed over a wide area, especially by hand.

BUD A small protuberance on a stem or branch, containing an undeveloped *shoot,* leaf, or flower.

BULBIL A small bulb-like part growing above ground on a flower stalk or in a leaf *axil.*

CAMBIUM A layer of cells in the stems and roots of vascular plants.

CHROMOSOME A body of the cell nuclei of plants, responsible for the determination and transmission of hereditary characteristics.

CLEFT A method of *grafting. Stock* is cut squarely across, split in the end, and one or two *scions* are inserted in the split so that the *cambiums* of stock and scion are in contact.

COLD FRAME A structure consisting of a wooden frame and a glass top, used for protecting young plants from the cold.

CORM An underground stem.

COTYLEDON The first or one of the first leaves to appear from a sprouting plant. Also called *true leaf.*

CROSS POLLINATION The transfer of pollen from one flower to the *stigma of another.*

CROWN The part of a plant, usually at ground level, between the roots and stem.

CUTTING A leaf, root, or stem, etc. of a plant, removed to form roots and to propagate a new plant.

DAMPING-OFF A disease of planted seeds or very young seedlings caused by fungi and resulting in death.

DECIDUOUS Shedding or losing foliage at the end of the growing season.

DIVISION A method of plant *propagation.* Parts are divided and segments capable of producing roots and shoots are planted. Parts of plants that can be used include *runners, crowns, tubers, and offsets.*

DRILL A furrow or trench in which seeds are planted.

EMBRYO The rudimentary plant contained within a seed.

ENDOSPERM The nutritive tissue of a plant seed, surrounding and absorbed by the *embryo.*

FLAT A shallow frame or box for seeds or seedlings.

FRIABLE With regard to soil, a soil that is crumbly, brittle.

FROND A large compound leaf of certain plants like palms and ferns.

FUNGICIDE A substance that destroys or inhibits the growth of fungi.

GENE A living organism that determines the size, growth, color, etc. of flowers. A gene is either male or female; genes usually occur in pairs.

GERMINATION The beginning of growth of a seed; the development of a *bud;* the production of a pollen tube; etc.

GRAFTING A *propagation* process. The joining together of parts of two different plants so they unite and grow as one plant. Methods of grafting include root, crown, top, *side,* whip and tongue, and *cleft.*

GREENWOOD See *softwood.*

HARDENING-OFF The appearance of dark green leaves on a plant.

HILL A circle of seeds, approximately 12 inches in diameter.

HOTBED A glass-covered bed of soil heated with electricity, used for the *germination* of seeds or for protection of tender plants.

HYBRID The offspring of genetically dissimilar parents or stock.

HYBRIDIZATION The transfer of pollen from one plant to another.

LAYERING The process of rooting branches, twigs, *shoots,* or stems that are still attached to a parent plant. The new growth is then cut and grown on its own. Methods of layering include *air, mound,* simple, and *tip.*

LEAF MOLD Decomposed leaves and other organic material used as humus or compost.

MARCOTTAGE An ancient Chinese method of *air layering.* The rooting medium is bound to the plant rather than enclosed in a pot or other container.

MOUND LAYERING A method of *propagation* in which various woody-stemmed plants are cut back to the ground in early spring and the new shoots that they develop are covered with soil to induce root growth. This root growth forms individual plants that can be removed in the fall.

MULCH A protective covering placed around plants to prevent evaporation of moisture and freezing of plants.

NEMATODE A plant-damaging worm.

NODE The often enlarged point on a stem where a leaf, *bud,* or other organ diverges from the stem to which it is attached; a joint.

OFFSET A shoot that develops laterally at the base of a parent plant, often rooting to form a new plant.

OVARY The part of a *pistil* containing the ovules.

PEAT The partly carbonized remains of vegetable tissue used as a *mulch* and plant food.

PERENNIAL A plant having a life span of more than 2 years.

PERLITE A natural volcanic glass used as a growing medium.

PETIOLE The stalk by which a leaf is attached to a stem.

PHOTO-PERIOD The amount of time that a plant is exposed to daylight.

PH SCALE A measure of the acidity or alkalinity of soil.

PISTIL The female, seed-bearing organ of a flower. It includes a *stigma* and *ovary.*

PROPAGATION The reproduction of a plant.

PROTHALLIA A small flat mass of tissue produced by a germinating spore of ferns, mosses, and related plants. Bears sexual organs and develops into a mature plant.

PRUNING The removing or cutting off of dead or alive plant parts to improve the shape or growth.

RHIZOME A root-like stem growing under or along the ground.

ROOTING HORMONE A synthetic plant hormone used to stimulate the growth of roots or *cuttings*.

RUNNER A slender creeping stem that puts forth roots from which *nodes* spaced at intervals along its length appear.

SCARIFICATION The process of nicking the outer coat of a seed to speed *germination.*

SCION A detached *shoot* or twig containing *buds* from a woody plant.

SEXUAL PROPAGATION The reproduction of a plant from seed.

SHOOT A stem with its leaves and other appendages.

SIDE GRAFT The *scion* is inserted into the side of the *stock* and the aerial head of the stock is permitted to grow until union is established between stock and scion.

SOFTWOOD The wood of a coniferous tree; also called *greenwood.*

SPHAGNUM Decomposed remains of sphagnum mosses.

SPORE An asexual reproductive organ characteristic of nonflowering plants.

STAMEN The male, pollen-producing reproductive organ of a flower.

STIGMA The top of the *pistil* of a flower upon which pollen is deposited at pollination.

STOCK A plant or stem onto which a *graft* is made; also a plant or tree from which *cuttings are taken.*

STOLON A stem growing along or under the ground and taking root at the *nodes* or apex to form new plants.

STRATIFICATION The placing of seeds in damp sand, peat moss, etc. to facilitate *germination.* Necessary for plants that need moisture or low temperature or both during their resting period.

SUCKER A secondary shoot.

TIP LAYERING The *propagation* of plants by bending a stem to the ground and covering the tip with soil so that roots and new shoots may develop.

TRENCH LAYERING A branch of a mature plant is grown in a horizontal position in a soil-filled trench. Soil is added periodically to cover the base of forming shoots. Plantlets are cut from the original stock at the end of the growing season and treated as new plants.

TRUE LEAF See *cotyledon.*

TUBER A swollen underground stem.

UNDERSTOCK The parent plant.

VERMICULITE Expanded mica used as a moisture-retaining growing medium.

VIABILITY The ability of a plant to live, develop, or germinate under favorable conditions.

# *Bibliography*

*Handbook on Propagation*, Brooklyn Botanic Gardens, Volume 13, No. 2, Sixth Printing, August 1970.

*Plant Propagation*, John P. Mahlstede and Ernest Haber, John Wiley & Sons, Inc., New York, New York, 1957.

*Plant Propagation in Pictures*, Montague Free, Doubleday & Company, Garden City, New York, 1957.

*Plant Propagation Practice*, James S. Wells, Macmillan Co., London, 1955.

*Plant Propagation: Principles and Practices*, Hudson T. Hartman and Dale E. Kester, Prentice-Hall, Englewood Cliffs, New Jersey, 1959.

*Practical Plant Propagation*, Alfred C. Hottes, A. T. De LaMare Co., Inc., New York, New York, 1925.

*Simple Practical Hybridizing for Beginners*, D. Gourley Thomas, John Gifford Ltd., Charing Cross Road, W.C., 1957.

*The Complete Book of Growing Plants from Seed*, Elda Haring, Hawthorn Books, Inc., New York, New York, 1967.

*The Genetics of Garden Plants*, M. B. Crane and W. J. C. Lawrence, Macmillan and Company, London, 1947.

# *Sources of Supplies*

Burgess Seed and Plant Co., Box 1140, Galesburg, Michigan 49053

Burpee Seed Co., 370 Burpee Building, Philadelphia, Pennsylvania 19132; Clinton, Iowa 52733; Riverside, California 92502

Henry Field Seed & Nursery Co., 19 North 12th Street, Shenandoah, Iowa 51601

Musser Forests, Box 73, Indiana, Pennsylvania 15701

Knaphill Nursery Ltd., Lower Knaphill, Woking, Surrey, England

George W. Park Seed Co., Inc., Greenwood, South Carolina 29547

Pearce Seed Co., Moorestown, New Jersey 08057

Clyde Robin, Castro Valley, P. O. Box 2091, California 94546

Henry Saier, Dimondale, Michigan 48821

R. H. Shumway, Rockford, Illinois 61101

Stoke's Seeds, Inc., 86 Exchange Street, Buffalo, New York 14205